Where Do Numbers Come From?

Why do we need the real numbers? How should we construct them? These questions arose in the nineteenth century, along with the ideas and techniques needed to address them. Nowadays it is commonplace for apprentice mathematicians to hear 'we shall assume the standard properties of the real numbers' as part of their training. But exactly what are those properties? And why can we assume them?

This book is clearly and entertainingly written for those students, with historical asides and exercises to foster understanding. Starting with the natural (counting) numbers and then looking at the rational numbers (fractions) and negative numbers, the author builds to a careful construction of the real numbers followed by the complex numbers, leaving the reader fully equipped with all the number systems required by modern mathematical analysis. Additional chapters on polynomials and quaternions provide further context for any reader wanting to delve deeper.

T. W. KÖRNER is Emeritus Professor of Fourier Analysis at the University of Cambridge. His previous books include *The Pleasures of Counting* and *Fourier Analysis*.

Where Do Numbers Come From?

T. W. KÖRNER
University of Cambridge

CAMBRIDGE
UNIVERSITY PRESS

CAMBRIDGE
UNIVERSITY PRESS

University Printing House, Cambridge CB2 8BS, United Kingdom

One Liberty Plaza, 20th Floor, New York, NY 10006, USA

477 Williamstown Road, Port Melbourne, VIC 3207, Australia

314–321, 3rd Floor, Plot 3, Splendor Forum, Jasola District Centre, New Delhi – 110025, India

79 Anson Road, #06–04/06, Singapore 079906

Cambridge University Press is part of the University of Cambridge.

It furthers the University's mission by disseminating knowledge in the pursuit of education, learning, and research at the highest international levels of excellence.

www.cambridge.org
Information on this title: www.cambridge.org/9781108488068
DOI: 10.1017/9781108768863

© T. W. Körner 2020

First published 2020

A catalogue record for this publication is available from the British Library.

Library of Congress Cataloging-in-Publication Data
Names: Korner, T. W. (Thomas William), 1946– author.
Title: Where do numbers come from? / T.W. Korner (University of Cambridge).
Description: Cambridge ; New York, NY : Cambridge University Press, [2020]
Identifiers: LCCN 2019020770 | ISBN 9781108488068
Subjects: LCSH: Number theory. | Mathematics – Philosophy.
Classification: LCC QA241 .K6697 2020 | DDC 512.7–dc23
LC record available at https://lccn.loc.gov/2019020770

ISBN 978-1-108-48806-8 Hardback
ISBN 978-1-108-73838-5 Paperback

Additional resources for this publication at www.cambridge.org/9781108488068.

Senseless as beasts, I gave men sense, possessed them
Of mind. I speak not in contempt of man;
I do but tell of good gifts I conferred.
In the beginning, seeing they saw amiss,
And hearing heard not, but, like phantoms huddled
In dreams, the perplexed story of their days
Confounded; knowing neither timber-work
Nor brick-built dwellings basking in the light,
But dug for themselves holes, wherein like ants,
That hardly may contend against a breath,
They dwelt in burrows of their unsunned caves.
Neither of winter's cold had they fixed sign,
Nor of the spring when she comes decked with flowers,
Nor yet of summer's heat with melting fruits
Sure token: but utterly without knowledge
Moiled, until I the rising of the stars
Showed them, and when they set, though much obscure.
Moreover, number, the most excellent
Of all inventions, I for them devised,
And gave them writing that retaineth all,
The serviceable mother of the Muse.

> *Aeschylus,* Prometheus Bound, *translation by G. M. Cookson*

What would life be without arithmetic but a scene of horrors.

> *Sydney Smith, letter to Miss Lucie Austin*

God made the integers, all else is the work of man. (Die ganzen Zahlen
hat der liebe Gott gemacht, alles andere ist Menschenwerk.)

> *Kroneckecker, reported by Weber,* Jahresbericht der Deutschen
> Mathematiker-Vereinigung *(1893)*

'When I use a word,' Humpty Dumpty said in rather a scornful tone, 'it
means just what I choose it to mean – neither more nor less.'

'The question is,' said Alice, 'whether you can make words mean so many
different things.' 'The question is,' said Humpty Dumpty, 'which is to be
master – that is all.'

> *Lewis Carroll,* Alice through the Looking-Glass

We should never forget that the functions, like all mathematical constructions, are only our own creations, and that when the definition, from which
one begins, ceases to make sense, one should not ask: what is it, but what

is it convenient to assume so that I can always remain consistent. Thus for example, the product of minus by minus.

Carl Friedrich Gauss, letter to Friedrich Bessel, 1811,
Volume 10 of his collected works

I have learnt one thing from my Arab masters, with reason as guide, but you another [from your teachers in Paris]: you follow a halter, being enthralled by the picture of authority. For what else can authority be called other than a halter? As brute animals are led wherever one pleases by a halter, but do not know where or why they are being led, and only follow the rope by which they are pulled along, so the authority of written words leads many people into danger, since they just accept what they are told, without question. So what is the point of having a brain, if one does not think for oneself?

Adelard of Bath, Conversations with His Nephew *(Adelard was one of those who introduced the Indian system of writing numbers to Europe.)*

Now you may ask, 'What is mathematics doing in a physics lecture?' We have several possible excuses: first, of course, mathematics is an important tool, but that would only excuse us for giving the formula in two minutes. On the other hand, in theoretical physics we discover that all our laws can be written in mathematical form; and that this has a certain simplicity and beauty about it. So, ultimately, in order to understand nature it may be necessary to have a deeper understanding of mathematical relationships. But the real reason is that the subject is enjoyable, and although we humans cut nature up in different ways, ... we should take our intellectual pleasures where we find them.

Richard Feynman, Addition and Multiplication, *Section 22-1 of the*
Feynman Lectures of Physics, *Volume 1*

The very important part played by calculation in modern mathematics and physics has led to the popular idea of a mathematician as a calculator, far more expert, indeed, than any banker's clerk, but, of course, immeasurably inferior, both in resources and accuracy, to what the 'analytic engine' will be, if the late Mr Babbage's design should ever be carried into execution.

But although much of the routine work of a mathematician is calculation, his proper work – that which constitutes him a mathematician – is the invention of methods.

Clerk Maxwell, review of Kelland and Tait's
Introduction to Quaternions *in* Nature, *1873*

There is no excellent beauty that hath not some strangeness in the proportion.

Francis Bacon, Essays

Have nothing in your houses that you do not know to be useful, or believe to be beautiful.

William Morris, Hopes and Fears for Art

Mathematical rigour is very simple. It consists in affirming true statements and in not affirming what is not true. It does not consist in affirming every truth possible.

Giuseppe Peano, quoted in Dictionary of Scientific Biography

There are still people who live in the presence of a perpetual miracle and are not astonished by it.

Henri Poincaré, The Value of Science

It seems to me, that the only objects of the abstract sciences or of demonstration are quantity and number, and that all attempts to extend this more perfect species of knowledge beyond these bounds are mere sophistry and illusion. As the component parts of quantity and number are entirely similar, their relations become intricate and involved; and nothing can be more curious, as well as useful, than to trace, by a variety of mediums, their equality or inequality, through their different appearances.

David Hume, An Enquiry Concerning Human Understanding

Contents

Appendix A   **Products of Many Elements**   233

Appendix B   **nth Complex Roots**   239

Appendix C   **How Do Quaternions Represent Rotations?**   243

Appendix D   **Why Are the Quaternions So Special?**   247

        References   255
        Index   257

Introduction

Faced with questions like 'What is truth?' or 'What is justice?', practical people dismiss them as useless speculation, and intellectuals enjoy the vague contemplation of matters beyond the reach of practical people. The best philosophers provide answers which may not be final, but which illuminate the paths that we follow.

The question 'What are numbers?' is clearly less important, but has interested several important philosophers and mathematicians. The answer given in this book is essentially that given by Dedekind in two essays, 'Stetigkeit und irrationale Zahlen' ('Continuity and irrational numbers') and 'Was sind und was sollen die Zahlen?' ('What are numbers and what should they be?'[1]) [7].

Starting with the natural numbers $\mathbb{N}^+$ (that is to say, the strictly positive integers), we construct the strictly positive rational numbers $\mathbb{Q}^+$ and then use these to construct the rational numbers $\mathbb{Q}$. We then use the rational numbers to construct the real numbers $\mathbb{R}$. Once $\mathbb{R}$ has been constructed, we construct the complex numbers $\mathbb{C}$ and we have all the numbers required by modern analysis (that is to say, calculus).

Of course, we still have to say where $\mathbb{N}^+$ comes from. Dedekind showed that all the properties of the strictly positive integers can be derived from a very small number of very plausible rules. The question of whether to accept these rules can now be left to the individual mathematician.

In the real world, we dig the foundations before we start the building. Historically, mathematicians have tended to install foundations when the building is half completed. Pedagogically, there are good reasons for studying the construction of the various number systems only after the student has acquired facility in using them.

---

[1] Another possible translation is 'What are numbers and what are they playing at?'

However, there will always be apprentice mathematicians who, when told 'we shall assume the standard properties of the real numbers', demand to know what these 'standard properties' are and why we can assume them. This book is written for those students and, for that reason, I have tried to follow an arc going from the relatively easy to the relatively difficult.

After a general discussion setting out some of the properties of the natural numbers, I show how to construct the strictly positive rational numbers (possibly better known to the reader as the positive fractions). By modifying the ideas involved, I then show how to construct negative numbers so that we get all the rationals.

We then take a long detour during which I discuss induction and modular arithmetic. The ideas are, I think, interesting in themselves and provide a background to the rather abstract arguments we then use to show how the properties of the natural numbers can be derived from a small number of postulates (the Peano axioms).

The final part of the book deals with the construction of the reals from the rationals and involves a substantial increase in the level of difficulty. Both the question of why we need the real numbers and the techniques used to construct them require ideas and techniques which only emerged, slowly and painfully, during the course of the nineteenth century. The easy construction of the complex numbers completes the task we have set ourselves.

There are two further chapters, one on polynomials and their roots and one on quaternions. They do not form part of the main argument, but, I think, help place it in context. The appendices are just that; the book is complete without them. The first two of the appendices are included for completeness rather than utility, the third requires a first-year university course in vector methods and the fourth a first-year university course in abstract algebra.

Because of the increasing difficulty of the discussion, the reader should not be surprised or disappointed if she can deal with only part of the book. I hope that she will come back later when she has gained more experience and read further.

If the reader finds this text too verbose or insufficiently precise, she should read Landau's *Foundations of Analysis* [20], a little gem of a book which covers the same ground with great precision[2] and without a wasted word. If she becomes fed up with my homespun philosophy, she will find it perfectly satirised in Linderholm's *Mathematics Made Difficult* [21].

---

[2] In 1977, Van Benthem Jutting wrote a PhD thesis in which the proofs in Landau's book were rewritten in an appropriate computer language and then computer checked [25]. According to Littlewood, Landau read proof sheets seven times, once for each of a particular type of error [22].

There are purported historical remarks scattered throughout this book. The reader will be protected from their worst effects if she constantly repeats the mantra 'But things were more complicated than that'.

Since I have tried to cater for relatively inexperienced mathematicians, readers who have done a year or more of university courses will find many things that they know: equivalence classes, isomorphism, induction, countability... (though sometimes viewed from a different angle). They should skim through those parts that are already familiar.

Some readers (probably the majority) will read this book to get a general idea of what is involved in setting up the various number systems. They should *read* the statements of the exercises carefully, but *work through* only the ones that interest them. Others will be less inclined to take the statements made in the exercises for granted. They should note that working things out for oneself leads to greater understanding, but, in case they cannot do an exercise, or they believe that it is the writer rather than the reader who should do the work, they will find sketch solutions to most of the exercises on my home page at

www.dpmms.cam.ac.uk/~twk/

together with a list of corrections.[3]

Mathematicians do not study the construction of number systems to serve any practical end. Reading this book will not make the standard university mathematics course any easier (except in so far as practising the art of proof in one branch of mathematics makes it easier to practise the art of proof in another). The reader must enjoy the contents of this book for their own sake. If she does, I shall consider myself well rewarded.

The reader should share my gratitude to four anonymous referees, several keen-eyed undergraduates and other readers and, in particular, Mr Guus Fabius for removing many errors and infelicities. I dedicate this book to two splendid daughters-in-law who have recently been kind enough to join my family.

---

[3] I will be very grateful to any reader who sends suggested corrections to my e-mail address, twk@dpmms.cam.ac.uk.

PART I

The Rationals

1

Counting Sheep

1.1 A Foundation Myth

The main bridge of the Finnish town of Tampere is decorated with four bronze statues representing the hunter, the maiden, the merchant and the tax gatherer. If we replace the maiden with the farmer, we obtain a plausible early history of the concept of number.

The early hunter needs only a few numbers. It may be useful to distinguish between one tiger and two, but, if there are more than two, counting is rarely relevant.[1] It is better to bring home three carcasses of deer rather than one, but, although it is good to know whether a bush has a few berries, quite a lot of berries or is loaded with berries, it is more useful to eat the berries than to count them.

The farmer lives in a settled state. He may, perhaps, know the names of every one of his sheep, but he will find it quicker to count his sheep to see if one is missing.[2] He will certainly count the apples or loaves he has in store, rather than giving them names. Sowing and shearing need to be done at particular times of the year, so the farmer requires some method of marking the progress of the days and counting provides a good way of doing this.

Although he does not require it, the farmer may find it useful to memorise some simple 'addition sums'. If he introduces three extra sheep into a field which already contains four, there will be seven sheep in the field. If he puts three apples into a box which already contains four, there will be seven apples

---

[1] It appears to be accepted that the Pirahã tribe of the upper Amazon use a 'one-two-many' system. (See [10].)

[2] To be more specific, *counting* produces a *number*. If a sheep is missing, then recounting will produce a different number. We shall be concerned mainly with number, but we shall look at the notion of counting in Section 6.4.

7

in the box. Here he is using the fact that sheep-counting numbers[3] behave like apple-counting numbers. A very occasional, very unusual, farmer may find this fact intriguing and justify it by thinking of each apple as being attached to a particular sheep. If he introduces three extra sheep bearing apples into a field which already contains four, there will be seven sheep bearing apples and so seven apples in the field.

A still more extraordinary farmer may note that, if he puts four sheep into a field which already contains three sheep, he still gets seven and would justify his observation with the following mental image:

● ● ● ● | ● ● ●=● ● ● ● ● ● ●=● ● ● | ● ● ● ●.

In modern notation, we write down the general law

$$a + b = b + a$$

and call it the *commutative law of addition*.

Note that if I put my shoes on before my socks, the result is not the same as if I put my socks on before my shoes. Not everything in this world is commutative.

Although the farmer may be self-sufficient in principle, there will be things that he desires, either for utility or for the sheer pleasure of ownership, which he cannot produce himself. Some of these things he can obtain by exchange with his neighbours, but other things come only from distant lands. Even in the Neolithic Age, coracles crossed the English Channel and Irish Sea carrying stone axes. The merchant will exchange so many stone axes for so many fleeces or, in later ages, such and such a weight of spices for a certain number of gold coins.

Times change and, in due course, complex societies began to emerge in Egypt, Babylonia and elsewhere in the Middle East. At its simplest, civilisation consists of one group of people with swords claiming to defend a group of farmers against another group of people with swords, but gathering people into cities and then into empires meant that they could build and manage complex irrigation schemes, build ports for trade and worship the gods in the style in which the gods wished to be worshipped (and which god would not prefer a massive temple complex with professional priests to the occasional amateur sacrifice of a sheep?). These things have to be paid for through

---

[3] Real life is always more complicated than theory. Shepherds in the Lake District of the UK use different numbers to count sheep 'Yain, Tain, Edderoa, ...' from those used to count other things, but they appear to be using a relic of some previous language rather than making a philosophical point.

taxes (possibly in the form of days of labour) and overseen by professional administrators.

In ancient Egypt, the administrative body was formed by scribes who could write and calculate. Here is an extract from a piece used for writing practice:

> See for yourself with your own eye. The occupations lie before you. The washerman's day is going up, going down. All his limbs are weak, [from] whitening his neighbours' clothes every day, from washing their linen. The maker of pots is smeared with soil, like one whose relations have died. His hands, his feet are full of clay; he is like one who lives in the bog. The cobbler mingles with vats. His odour is penetrating. His hands are red with madder, like one who is smeared with blood. He looks behind him for the kite, like one whose flesh is exposed. The watchman prepares garlands and polishes vase-stands. He spends a night of toil just as one on whom the sun shines. The merchants travel downstream and upstream. They are as busy as can be, carrying goods from one town to another. They supply him who has wants. But the tax collectors carry off the gold, that most precious of metals. The ships' crews from every house (of commerce), they receive their loads. They depart from Egypt for Syria, and each man's god is with him. (But) not one of them says: 'We shall see Egypt again!' The carpenter who is in the shipyard carries the timber and stacks it. If he gives today the output of yesterday, woe to his limbs! The shipwright stands behind him to tell him evil things. His outworker who is in the fields, his is the toughest of all the jobs. He spends the day loaded with his tools, tied to his tool-box. When he returns home at night, he is loaded with the tool-box and the timbers, his drinking mug, and his whetstones. The scribe, he alone, records the output of all of them. Take note of it!
>
> Papyrus Lansing, *English translation (slightly altered)*
> *by A. M. Blackman and T. E. Peet*

Some suggestions of the kind of task a scribe might expect are given in a long sarcastic letter berating an incompetent pupil (or perhaps colleague). I give two extracts.

> There is to be constructed a ramp of 730 cubits in length with a width of 55 cubits, containing 120 compartments provided with rushes and beams, having a height of 60 cubits at its summit and 30 cubits at its middle, with a batter of 15 cubits, while its base is of 5 cubits. The amount of bricks required for it is asked of the commander of the workforce. The scribes are all gathered together through lack of one who knows among them. So they all put their trust in you, saying, 'You are an expert scribe, my friend. Decide for us quickly. See, your name is celebrated. Let one be found in this place capable of magnifying the other thirty. Don't let it be said of you that there is anything of which you are ignorant. Answer for us the amount of bricks required for it. Look, its dimensions(?) are before you with each one of its compartments being 30 cubits long and 7 cubits wide.'...
> ...O you scribe, so alert and competent that there is nothing at all of which you are ignorant, who blazes in the darkness at the head of the troops and illumines for them, you are dispatched on a mission to Djahy at the head of the victorious army

in order to crush those rebels who are called Naarin-warriors. The host of soldiers that is under your charge comprises 1 900 Egyptians, 520 Sherden, 1 600 Kehek, [100] Meshwesh and 880 Nubians, a total of 5 000 all told, apart from their captains. There are brought to you bonus rations into your presence: bread, sheep and goats, and wine. The number of men is too large for you, and the foodstuff is insufficient for them: only 300 sweet loaves, 1 800 cakes, 120 assorted goats and sheep, and 30 jugs of wine. The army is so numerous that the foodstuff has been under-estimated as though you had pilfered from it. You receive it in charge to be deposited in the camp. The troops are prepared and ready, so divide it quickly into portions, each man's share into his hands . . . Midday is come, and the camp is hot. One says, 'It is time to move on. Do not make the troop marshaler angry. We still have a long march ahead of us. What bread do we have at all? Our night camp is far away. O What's-your-name?, what's the sense of scourging us so, when you are supposed to be an expert scribe?' It is only after six hours have elapsed in the day that you proceed to distribute the provisions.

Papyrus Anastasi, *English translation (slightly altered) by E. F. Wente*

I may be mistaken if, across thousands of years, I sense a delight in computation for its own sake. Certainly we see how a more complex society requires more complex arithmetic and, in particular, the operation of multiplication. Returning to mathematics as we see it, rather than as our remote forebears may have seen it, we shall try to find some of the general laws governing this operation. We have seen that $a+b = b+a$. Multiplication follows a similar law,

$$a \times b = b \times a$$

the *commutative law of multiplication*, illustrated by the following diagram in which we count the dots in a rectangle and its rotation to obtain $3 \times 4 = 4 \times 3$:

We may also remember the pleasure that reciting the 'one times' table gave us in our youth and add

$$1 \times a = a.$$

We have some further observations, which would hardly trouble an Egyptian scribe working with a clear sequence of calculations, but which are clearly relevant to anyone who has used an electronic calculator. These concern bracketing – that is to say, specifying the order in which a set of calculations is to be performed. The first law is the *distributive law*:

$$a \times (b + c) = (a \times b) + (a \times c).$$

The statement of the distributive law can be simplified by applying the standard convention that multiplication has priority over addition unless prevented by bracketing.[4] We then have

$$a \times (b + c) = a \times b + a \times c,$$

illustrated by the following diagram for the calculation $3 \times (2 + 4) = 3 \times 2 + 3 \times 4$:

$$
\begin{array}{ccc}
\bullet\ \bullet\ | & \bullet\ \bullet\ \bullet\ \bullet & \bullet\ \bullet\ \bullet\ \bullet\ \bullet \\
\bullet\ \bullet\ | & \bullet\ \bullet\ \bullet\ \bullet\ = \bullet & \bullet + \bullet\ \bullet\ \bullet\ \bullet\ . \\
\bullet\ \bullet\ | & \bullet\ \bullet\ \bullet\ \bullet & \bullet\ \bullet\ \bullet\ \bullet\ \bullet
\end{array}
$$

Exercise 1.1.1 *Use the distributive law just stated together with the commutative law of multiplication to show that*

$$(b + c) \times a = (b \times a) + (c \times a).$$

The next rule, (the *associative law of addition*),

$$a + (b + c) = (a + b) + c,$$

would appear obvious to any ancient farmer. If we have three adjoining pens containing sheep and remove the hurdles between the first two pens and then remove the remaining hurdles we will have the same number of sheep in the enlarged pen as if we had removed the hurdles in the opposite order. We illustrate the case $2 + (3 + 4) = (2 + 3) + 4$:

$$\bullet\bullet|(\bullet\bullet\bullet|\bullet\bullet\bullet\bullet) = \bullet\bullet|\bullet\bullet\bullet\bullet\bullet\bullet\bullet = \bullet\bullet\bullet\bullet\bullet\bullet\bullet\bullet\bullet = \bullet\bullet\bullet\bullet\bullet|\bullet\bullet\bullet\bullet = (\bullet\bullet|\bullet\bullet\bullet)|\bullet\bullet\bullet\bullet.$$

However, a chemist would not expect that mixing chlorine with sodium and then mixing the result with water would have the same effect as first mixing sodium with water and then mixing chlorine with the result.[5]

Exercise 1.1.2 *Let us define an operation $\boxplus$ by $a \boxplus b = 2 \times (a + b)$. Verify the following results.*

(i) $a \boxplus b = b \boxplus a$.
(ii) $a \times (b \boxplus c) = (a \times b) \boxplus (a \times c)$.
(iii) For some choice of a, b and c, to be given explicitly,

$$a \boxplus (b \boxplus c) \neq (a \boxplus b) \boxplus c.$$

---

[4] This book is heavily 'over-bracketed', but, where the reader finds the bracketing insufficient, she is invited to use the standard rules.
[5] Do not try this at home.

Finally, we have an *associative law for multiplication*, analogous to the associative law of addition:

$$a \times (b \times c) = (a \times b) \times c.$$

Our putative ancient farmer might scratch his head at the idea that it would ever be necessary to multiply *three* numbers together. The scribe, accustomed to dealing with large quantities of identical bricks, would observe that to obtain a stack of bricks ten bricks in length, twelve bricks in width and five bricks in height we could either build ten neighbouring walls twelve bricks in width and five bricks in height or layer floors of bricks ten bricks in length and twelve bricks in width five times.

1.2 What Were Numbers Used For?

If the reader would like a proper account of ancient Egyptian mathematics, she could read Imhausen's excellent book [15], noting her comment that 'The amount of available literature on Egyptian mathematics is even more astonishing if we take into account that the only sources on which almost all these studies were founded are four papyri (of which half consist only of a number of fragments), a leather roll, a wooden board and two ostraca'.[6]

However, since we are considering what might have been, rather than what was, we can speculate about how ancient civilisations used numbers. If we think about tax collection, palace organisation or collecting port dues, we see that the main use of numbers would not be for calculation, but for record keeping. Record keeping requires, above all, consistency and standardisation. It is very difficult to marry innovation with consistency and standardisation,[7] so there were excellent reasons why societies would stay with whatever methods of writing numbers and whatever associated methods of calculation they inherited from the previous generation. If we consider, in addition, mankind's cultural conservatism,[8] we see that the correct question to ask in relation to the history of mathematics (at least up to AD 1700) is not 'Why did things stay the same?' but 'Why did things change?'

Of course, the use of numbers could not be confined to simple record keeping. A temple would wish to keep a record of individual donations, but would

---

[6] Stone or pottery shards used as writing material.

[7] In modern times, important information held in obsolete electronic formats may be lost, or at least very difficult to recover.

[8] A good example of cultural conservatism is given by my internal numbering for exercises and theorems (see, for example, Theorem 2.3.1).

also need to know the total sum produced. Temple scribes would thus need to perform addition. If, as the Egyptians did, we count in units, tens, hundreds and so on this is easily done using some version of an abacus. At its simplest, this consists of shallow depressions in the ground containing pebbles. Thus we might have an arrangement

$$\left| \begin{array}{c|c|c} C & B & A \\ \bullet\bullet\bullet\bullet & \bullet\bullet & \bullet\bullet\bullet \end{array} \right|$$

where the units hole A contains 3 pebbles, the tens hole B contains 2 pebbles and the hundreds hole C contains 4 pebbles, so we have the number four hundred and twenty three. Suppose that we wish to add this to four hundred and seven which we represent using pebbles as before:

$$\left| \begin{array}{c|c|c} C' & B' & A' \\ \bullet\bullet\bullet\bullet & & \bullet\bullet\bullet\bullet\bullet\bullet\bullet \end{array} \right|$$

(note that the tens hole B' is empty). We now just add the contents of hole A' to hole A, those of hole B' to hole B and so on subject to the additional rule (which will be familiar to you from school) that, whenever a hole contains ten or more pebbles, ten pebbles should be removed and one pebble placed in the next hole to the left. In the example given, we get

$$\left| \begin{array}{c|c|c} C & B & A \\ \bullet\bullet\bullet\bullet\bullet\bullet\bullet\bullet & \bullet\bullet\bullet & \end{array} \right|,$$

that is to say, 830.

It should be remarked that users of the modern counting frame abacus can add faster than modern users of paper and pencil.[9] The use of an abacus will seem less strange to my readers, who usually confide their calculations to a little electronic device, than it did to people of my generation who did their own pencil-and-paper calculations.

Multiplication is less useful for record keeping unless we are dealing with standardised objects, for example, calculating the number of bricks required for a wall or providing one day's ration for an army.

It is believed that the ancient Egyptians used a procedure for multiplication close to what is now called 'Russian peasant multiplication'. Given two numbers a and b with $a < b$, we place a in a first column α and b in a second

[9] I can bear personal witness to this, having given out the prizes at a splendid abacus demonstration by young school children in Zhenjiang.

column β and proceed as follows. Either a is even and we calculate $a' = a/2$ and $b' = b \times 2$, placing a' in column α and b' in column β, or a is odd and we calculate $a' = (a-1)/2$ and $b' = 2 \times b$, keeping our result in columns α and β as before, but keeping a note of b in a third column γ. We repeat this process with a' and b' and continue until we obtain the pair 1 and B, say. The product $a \times b$ equals B plus the sum of the values noted in the third column. Here is an example in which we use the method to multiply 27 by 14:

α	β	γ
14	27	
7	54	
3	108	54
1	216	108

We have $216 + 54 + 108 = 378$, and 378 is indeed the correct answer.

Exercise 1.2.1

(i) Compute 45×103 using Russian peasant multiplication.
(ii) Explain why Russian peasant multiplication works.

[*We shall give a couple of explanations in Exercise 4.3.15.*]

Simple division sums such as dividing twenty one loaves among seven workmen could be tackled by enlightened guess work (as indeed most people would do such sums today) but 'long division' must have been very difficult. There were also problems with what we now call fractions. Faced with the problem of dividing five loaves among six workmen, a natural procedure is to cut three loaves in half and then cut the remaining loaves into thirds. Each man would receive half a loaf and a third of a loaf. In modern terms,

$$\frac{5}{6} = \frac{1}{2} + \frac{1}{3}.$$

The ancient Egyptians extended this idea so a number x with $1 > x > 0$ would be represented in terms corresponding to the right-hand side of the equation

$$x = \frac{1}{n_1} + \frac{1}{n_2} + \cdots + \frac{1}{n_k},$$

with $2 \leq n_1 < n_2 < \cdots < n_k$. We call this an Egyptian fraction expansion.

Exercise 1.2.2

(i) Does there exist an Egyptian fraction expansion for all rational x with $1 > x > 0$?
(ii) Is it unique?

[Think about these questions, but do not worry if you cannot come to a conclusion. Part (ii) is answered in the next paragraph and part (i) in Exercise 4.3.14.]

Unfortunately, although 'Egyptian fractions' give a reasonable way of dividing loaves and recording what we would now call fractions with small denominators,[10] they seem deeply unfitted for calculation. How, for example, could one obtain the equality

$$\frac{1}{3} + \frac{1}{5} = \frac{1}{2} + \frac{1}{30}$$

working only in terms of 'Egyptian fractions'?

To what extent would these limitations have affected the Egyptians? Surely, the answer must be, hardly at all. It might have taken some time to work out the daily rations for a work party of eight men, but, once the calculation was made, it would apply to all work parties of eight men on all days. The construction of a pyramid would be a major administrative feat and would involve quite a lot of arithmetic (how much must be a matter of speculation), but the effort involved in arithmetical calculations must have been dwarfed by the other tasks involved. We do not even need to guess that the results of difficult calculations must have been tabulated for reuse, because one of the major pieces of mathematical evidence to have survived is a table of Egyptian fraction expansions for what we would now write as $2/(2n + 1)$.

With no pressure for change and no indication as to what change might be desirable, ancient societies continued with whatever arithmetical systems their forefathers had used.

1.3 A Greek Myth

As in earlier and later societies, most ancient Greeks spent the greater part of their time farming, trading and fighting one another. However, it was noted, both by those who admired their behaviour and by those who despised it, that some Greeks liked nothing better than abstract speculation and heated argument.

Faced with an account of number along the lines given in the previous sections, we can imagine one of these trouble-makers raising various objections.

---

[10] But not when the denominators are large. This would not matter for the kind of uses required by the ancients.

(1) You talk about the number of sheep in a flock as a well-defined unchanging object. But anyone who owns a large flock of sheep knows that he may go to sleep the owner of a hundred sheep and wake up the owner of a hundred and one or ninety nine sheep.

(2) You say that we can derive multiplication by looking at a merchant who charges two silver pieces for a roll of cloth and six pieces of silver for three rolls of cloth. But it is common for such a merchant to offer five rolls of cloth for eight pieces of silver. Should we conclude that $5 \times 2 \underset{?}{=} 8$?

(3) You say that the truth of the result $a + b = b + a$ can be grasped by dividing a line of dots in two different ways. I might agree (for the sake of argument) that we can see that $5 + 3 = 3 + 5$ in this way, but you claim that this result continues to hold for numbers as large as several million. I cannot picture a collection of several million dots and I do not think you can.

(4) You say that we can use the same numbers for counting sheep, goats or days. At a pinch, I will agree that adding three sheep to four goats produces seven animals, but I cannot see how to add three sheep to four days.

(5) Things are even worse when it comes to multiplication. I say that you cannot multiply three sheep by five sheep and you reply that you mean three flocks of five sheep – but if, as you claim, multiplication is commutative, why is the answer fifteen sheep rather than fifteen flocks?

So far as I know, my Greek trouble-maker is entirely imaginary, but we do know that the Greeks considered similar problems which arise in the study of space, that is to say, geometry.

One possible solution (but not theirs) to these problems is the following. By the study of the real world and by introspection, we can come up with a set of laws which appear to be followed, more or less, by the points, lines, lengths and angles of the real world. We then study what further laws can be strictly deduced from our initial set of laws. (In the language of mathematics what *theorems* can be deduced from our *axioms*.) We have made geometry a branch of pure mathematics (or in Hogben's hostile account, 'a respectable form of relaxation for the opulently idle' [14]).

It is then the job of the applied mathematician to decide under what circumstances and to what extent our abstract geometry can be applied to the imperfect points and lines of the real world.

Unimplemented, these ideas are just hot air. However, the 'Euclidean geometry' of the Greeks turned out to be one of the most beautiful and influential objects in the history of mankind. At the same time, the Greek thinkers showed that mathematics could be applied to the study of nature, including not only the movement of the stars (where the Babylonians and Egyptians had already made strides) but also things of this earth. Archimedes moved easily from the most abstract geometry to the study of the stability of ships.

Hardy in his *A Mathematician's Apology* [13], page 12, puts it very well:

> Greek mathematics is the real thing. The Greeks first spoke a language which modern mathematicians can understand; as Littlewood said to me once; they are not clever schoolboys, or 'scholarship candidates', but 'Fellows of another college'.[11]

Can we produce an axiomatic treatment for the natural numbers? Let us write down the rules we have obtained so far.

$$a + b = b + a \qquad \text{(Commutative law of addition)}$$
$$a \times b = b \times a \qquad \text{(Commutative law of multiplication)}$$
$$a + (b + c) = (a + b) + c \qquad \text{(Associative law of addition)}$$
$$a \times (b \times c) = (a \times b) \times c \qquad \text{(Associative law of multiplication)}$$
$$1 \times a = a \qquad \text{(One is a unit)}$$
$$a \times (b + c) = (a \times b) + (a \times c) \qquad \text{(Distributive law)}$$

We can stare at these rules for a long time while they stare back at us, but nothing happens. There is more to the axiomatic method than writing down a set of axioms and drawing appropriate consequences. Unless the axioms give *interesting* results deduced in an *interesting* way,[12] the system is not worth studying.

Exercise 1.3.1 *Use our axioms to show that*

$$a \times 1 = a.$$

The rules we have collected so far do not capture the nature of the natural number system. We shall see there are many systems that obey these rules. Here is a particularly simple one under a slight disguise (think of the rules for addition and multiplication of odd and even numbers). It consists of two elements θ and 1 with the rules

$$\theta + \theta = 1 + 1 = \theta, \ 1 + \theta = \theta + 1 = 1, \ \theta \times \theta = \theta \times 1 = 1 \times \theta = \theta, \ 1 \times 1 = 1.$$

---

[11] People from other universities may consider this a very Cambridge thing to say. People from other colleges consider this a very Trinity thing to say.

[12] Of course, not all of the results and not all of the deductions need be interesting – just some of them.

Exercise 1.3.2 *Check, to your own satisfaction, that the system does indeed obey all the stated rules.*
[This is a special case of the system $(\mathbb{Z}_p, +, \times)$ which will be discussed in Section 5.1, with $p = 2$.]

On a more practical level, it may occur to us that a hunter prefers to be chased by one tiger rather than two and a farmer prefers to have fifteen sheep rather than ten. We therefore introduce the idea of order $>$ linked to addition by the following definition which we shall call the *order rule*.

Definition 1.3.3 *Let a, $b \in \mathbb{N}^+$. We have $a > b$ if and only if we can find a natural number $c \in \mathbb{N}^+$ such that $b + c = a$.*

Lemma 1.3.4 *The order $>$ just defined on $\mathbb{N}^+$ is transitive, that is to say, if $a > b$ and $b > c$, then $a > c$.*

Proof. If $a > b$ and $b > c$, then, by definition, we can find u and v such that $a = b + u, b = c + v$. Using the associative law of addition we have

$$a = b + u = (c + v) + u = c + (v + u),$$

so $a > c$.   ∎

When we write $a > b > c$, we implicitly refer to transitivity.

In order to make full use of our order $>$ we need to introduce a new law which, although it may appear obvious to the tax collector, does not follow from the laws so far introduced.
Trichotomy.[13] The order $>$ just defined on $\mathbb{N}^+$ obeys trichotomy, that is to say, exactly one of the following conditions holds: $a > b$ or $b > a$ or $a = b$.

Exercise 1.3.5 *Consider the system of Exercise 1.3.2. Write $x \ominus y$ if we can find a u such that $x = y + u$. Check that $x \ominus y$ whichever x and y we choose. In particular, $\theta \ominus 1$ and $1 \ominus \theta$, so trichotomy fails.*

The idea of trichotomy will appear over and over again throughout this book.
We shall follow standard mathematical practice by writing $a \geq b$ if $a > b$ or $a = b$ while using $a < b$ to mean $b > a$ and $a \leq b$ to mean $b \geq a$.

Exercise 1.3.6 *Show that transitivity and trichotomy are together equivalent to the following rules:*

[13] In a dichotomy exactly one of two things is true; in a trichotomy exactly one of three things is true.

(1) If $a \geq b$ and $b \geq a$, then $a = b$.
(2) If $a \geq b$ and $b \geq c$, then $a \geq c$.
(3) At least one of the following conditions holds: $a \geq b$ or $b \geq a$.
(4) If $a > b$, then $a \neq b$.

[*In other words, show that rules (1), (2), (3) and (4) follow from transitivity and trichotomy and vice versa.*]

Exercise 1.3.7 *Explain why trichotomy enables us to define* max$\{a, b\}$ *(the maximum of a and b) and* min$\{a, b\}$ *(the minimum of a and b).*
   Show that

$$\max\{a, b\} + \min\{a, b\} = a + b.$$

Inequality interacts with our previously defined operations. In particular, we have two further results which we call the *addition and multiplication laws for inequalities*.

Lemma 1.3.8 *Consider order $>$ just defined on $\mathbb{N}^+$.*

(i) If $a > b$, then $a + c > b + c$.
(ii) If $a > b$, then $a \times c > b \times c$.

Proof. (i) If $a > b$, then we can find a u such that $a = b + u$. By the associative and commutative laws of addition

$$a + c = (b + u) + c = b + (u + c) = b + (c + u) = (b + c) + u,$$

so $a + c > b + c$.

   (ii) If $a > b$, then we can find a u such that $a = b + u$. By the distributive law and the commutative law of multiplication

$$a \times c = (b + u) \times c = c \times (b + u) = (c \times b) + (c \times u) = (b \times c) + (c \times u),$$

so $a \times c > b \times c$. ∎

   Combining these addition and multiplication laws for inequalities with trichotomy we get a number of useful *cancellation laws*.

Lemma 1.3.9 *If we work with the natural numbers $\mathbb{N}^+$, then the following results hold.*

(i) If $a + c = b + c$, then $a = b$.
(ii) If $a \times c = b \times c$, then $a = b$.
(iii) If $a + c > b + c$, then $a > b$.
(iv) If $a \times c > b \times c$, then $a > b$.

Proof. (i) By trichotomy we know that exactly one of the following holds: $a > b, b > a$ or $b = a$. If $a > b$, then $a + c > b + c$ and (by trichotomy again) $a + c \neq b + c$. If $b > a$, then $b + c > a + c$ and $a + c \neq b + c$. Thus, if $a + c = b + c$, we have $a = b$.

Results (ii), (iii) and (iv) are left as an exercise for the reader. ■

Exercise 1.3.10 *Prove Lemma 1.3.9 (ii), (iii) and (iv).*

Notice that the cancellation law for addition enables us to strengthen the order rule.

Lemma 1.3.11 *If $a > b$, there is a unique c such that $a = b + c$.*

Proof. Existence follows from the order rule. To prove uniqueness observe that if $a = b + c$ and $a = b + c'$, then $b + c = b + c'$ and Lemma 1.3.9 (i) shows us that $c = c'$. ■

For more than 2,000 years Euclidean geometry was the only example of an interesting system of complex logical deductions from apparently simple premises. As such it, rightly, enjoyed enormous prestige.

> In the course of my law reading I constantly came upon the word *demonstrate*. I thought, at first, that I understood its meaning, but soon became satisfied that I did not. I said to myself, What do I do when I *demonstrate* more than when I *reason* or *prove*? How does demonstration differ from any other proof? I consulted Webster's Dictionary. They told of 'certain proof,' 'proof beyond the possibility of doubt'; but I could form no idea of what sort of proof that was. I thought a great many things were proved beyond the possibility of doubt, without recourse to any such extraordinary process of reasoning as I understood *demonstration* to be. I consulted all the dictionaries and books of reference I could find, but with no better results. You might as well have defined *blue* to a blind man. At last I said,— Lincoln, you never can make a lawyer if you do not understand what *demonstrate* means; and I left my situation in Springfield, went home to my father's house, and stayed there till I could give any proposition in the six books of Euclid at sight. I then found out what *demonstrate* means, and went back to my law studies.
>
> Henry Ketcham, *The Life of Abraham Lincoln*

Versions of Euclid's geometry were often placed at or near the centre of the educational system. I belonged to the last generation in the UK for whom this was true and I vividly recall the pleasure the subject gave me.

Many attempts were made to produce axiomatic treatments of such subjects as philosophy, theology and politics but, although some were interesting, none were successful.[14]

---

[14] Indeed, up to now, even very mathematical parts of physics seem to resist interesting axiomatic development.

In the early nineteenth century, it was discovered that Euclid's axioms could be modified to give other geometries. Later, it was found that some of the new subjects which were being introduced into mathematics, such as group theory, could be given an axiomatic presentation. Problems with rigorising calculus led to the kind of axiomatic treatment of numbers considered in this book. We now have many examples of axiomatic systems within mathematics to study and compare.

Although the popular image of a mathematician as a lone worker is not entirely false, it is an oversimplification. Mathematical knowledge passes from teacher to pupil and from colleague to colleague. If the chain is broken, an exceptional mind is required to reforge it. If only because time for thought has to be paid for, mathematics will flourish more in a favourable than in an unfavourable society.

Over time, conditions became more unfavourable to the Greek mathematical school. Cicero says, "Geometry was in high esteem with [the Greeks], therefore none were more honourable than mathematicians. But we [the Romans] have confined this art to just measuring and calculating." The later Roman empire was beset by economic and political problems, and its dominant intellectual system was more concerned with personal salvation than with the subtleties of the Greeks.[15] For these, or, perhaps for other reasons, the direct Greek mathematical tradition came to an end.

---

[15] What has Athens to do with Jerusalem? (Tertullian)

2

The Strictly Positive Rationals

2.1 An Indian Legend

There are many ways of recording numbers. At the simplest, when counting sheep, we can just use one stroke to record each sheep so that |||| stands for four sheep. However, this is no good if the number is large; so, for example, the ancient Egyptians had a sign for ten and for one hundred and so on. If we take ∩ as our sign for ten and ∪ as our sign for one hundred, we can write three hundred and twenty four as ∪ ∪ ∪ ∩ ∩||||. Notice that this is reasonably brief and makes it easy to record numbers from and enter numbers into an abacus.[1]

The ancient Greeks used a more compact, but more complex, scheme, with different letters representing different multiples of ten, one hundred and so on. We can imagine a representing one, A representing ten, α representing one hundred, b representing two, B representing twenty, β representing two hundred, c representing three and so on. In this notation $\beta C a$ would be two hundred and thirty one.

The ancient Indians used a similar scheme, but someone at sometime, or several people over a length of time, introduced a clever modification which only used nine symbols (which we shall represent as 1, 2, ..., 9 with their usual meanings) and a placeholder *. The number written in Greek style as $\delta A c$ would be written as 413 where we can tell that 3 represents three because it is in the first place starting from the right, that 1 represents ten because it is in the second place and 4 represents four hundred because it is in the third place from the right.

This is clearly a more economical scheme, but how are we to represent the Greek number δc? The simple, but tremendous, answer is to use the

---

[1] The Roman system of writing I, II, III, IV, V, VI, VII, VIII, IX, X, XI ... achieves slightly greater brevity if you wish to record things, but makes computation harder.

placeholder ∗ to represent the empty second column and write 4 ∗ 3. Sir Thomas Browne tells us that 'What song the Sirens sang, or what name Achilles assumed when he hid himself among women, though puzzling questions, are not beyond all conjecture.' Scholars have debated whether the origins of the idea of a placeholder are to be found in the tables of ancient Babylonian astronomers or whether we should thank some humble anonymous Indian temple scribe.

Be that as it may, it is not immediately clear why the new system is substantially better than the others we have described. It allows us to represent very large numbers rather easily, but the ancient and medieval worlds did not need very large numbers and, had they, it would have been possible to modify the earlier systems to cope.[2] The new system is much less intuitive than the old. Anyone can understand that ∪ ∪ ∩ ∩ ∩|| represents two sticks, three bundles of ten sticks and two bundles of a hundred sticks, but it requires special training to recognise that 1 in the first place starting from the right in 121 magically becomes one hundred when placed in the third place.

Exercise 2.1.1 *We write our numbers with the size of bundles decreasing as we move from left to right so that 341 is three hundred and forty one. Spend a little time using the opposite convention. For example, using the reverse convention throughout, write down three hundred and seventy nine and two hundred and sixty five and then use long multiplication to multiply them together, finally translating the answer back into words.*

Exercise 2.1.2 *The translation from Indian notation to words is not quite as simple as my account implies. The reader is invited to translate*

$$2\,140\,676\,912\,926\,927$$

(a purported estimate of the US federal debt in cents at a particular instant) into words. She should then write down general instructions for doing such a translation.

---

[2] Archimedes actually discusses this problem in *The Sand Reckoner*. 'There are some, King Gelon,' says Archimedes 'who think that the number of grains of sand is infinite in multitude; and I mean by the sand not only that which exists about Syracuse and the rest of Sicily but also that which is found in every region whether inhabited or uninhabited. Again there are some who, without regarding it as infinite, yet think that no number has been named which is great enough to exceed its multitude. ... But I will try to show you by means of geometrical proofs, which you will be able to follow, that, of the numbers named by me and given in the work which I sent to Zeuxippus, some exceed not only the number of grains of sand [required to fill a sphere the size of the earth], but also the number of grains of sand [required to fill the universe theorised by Aristarchus].'

The new system was invented sometime between the first and fourth centuries in India and spread slowly. In the ninth century AD, the great Persian mathematician Al-Khwārizmī wrote *On Calculation with Hindu Numerals* which helped introduce the system to the Arab world and thence onward to Europe where, over the course of several centuries, it displaced Roman numerals.

The ultimate triumph of the Indian system is generally attributed to the fact that it allowed the invention of new methods of calculation. It was true that addition on an abacus remained faster than pen-and-paper addition with the new numerals, but the new methods of long multiplication and long division, made possible by the Indian system, brought previously extremely difficult operations within the grasp of ordinary people.[3]

Exercise 2.1.3 *(It is very likely that the contents of this exercise are familiar to the reader.)*

(i) *The design of the abacus and the pencil-and-paper methods we learn for calculating in school take into account the fact that we have ten digits and group things in ones, tens, hundreds and so on. If, like some characters in cartoon films, we had eight digits, we would count in ones, eights, sixty-fours and so on. Such a system (which we say has* radix 8*) would have the elementary symbols* 1, 2, 3, ..., 7 *together with the placeholder* 0*. Perform the addition and multiplication sums indicated, working with radix*

$$8, \quad \begin{array}{r} 153 \\ +672 \end{array} \quad and \quad \begin{array}{r} 53 \\ \times 72 \end{array}.$$

(ii) *Our machines like to work in radix 2, that is to say, binary. What binary expression corresponds to the radix 10 expression* 104*? What radix 10 expression corresponds to the binary expression* 10011*?*

[*From time to time in this book we shall refer to binary notation, so the reader should make sure she is happy with the idea.*]

Exercise 2.1.4 *It is natural that we should consider 'our' number system better than its predecessors, and the fact that it has been universally adopted is good evidence that we are right. What reasons, if any, do you have for believing that there does not exist one which is substantially better? (Changing the radix would certainly not justify the work involved.)*

---

[3] Notice how apparently unrelated advances reinforce one another. Paper-making, invented in China a thousand years before, greatly reduced the cost and greatly increased the availability of writing material in Europe, thus making 'pen-and-paper' methods practicable. Gutenberg's method of printing with movable type made manuals of arithmetic cheap and plentiful and so aided rapid transmission of the new ideas.

The placeholder (or 'zero' as we shall now call it) was not the only inno-vation of the new Indian arithmetic. It also introduced a new way of writing 'fractions' essentially equivalent to our present

$$\frac{\text{numerator}}{\text{denominator}}.$$

This can be contrasted, not only with Egyptian fractions, but also with the Roman system of using twelfths[4] and a collection of ad hoc smaller units such as forty-eighths, each with their own set of symbols.

Previous systems allowed fractions to be used in records, but the new Indian system allowed ordinary people to add and multiply fractions just as you learned in school. The instruction manuals for the new arithmetic made much of this new advance, but who (apart from puzzle lovers) would want it?

The answer returns us to the merchant or, more particularly, to a new class of merchants. Even when Baghdad and Constantinople looked with dis-dain at the uncivilised people of Europe, there were long-distance traders, but they practised the kind of trade in which you ended up rich or dead, and where a sharp sword (or retinue with sharp swords) was more useful than any but the simplest arithmetic. Eventually, Europe settled down into a collection of states, often quite small, each peaceful (more or less) within its own boundaries and each with its own army, weights and measures and currency.

Trade now became safer but also more competitive. A London money changer would need to know the value of a Mark of Lubeck, a Mark of Riga, a Mark Sundische, a Mark of Cologne, a Mark Wendische, a Pound of Vienna, a Groote of Holland, a Flemish Mouton, ... and so on. Moreover, since gov-ernments were always tempted to reduce the amount of precious metal in a coin, the same coin would have different values if issued at different dates. If he set his rates wrong, our money changer would lose out against more sophis-ticated competitors. When profits depend on fractional advantage, fractional calculations become essential.

As trade increased, so did its financial complexity. Traders wanted letters of credit. Ship owners sought insurance. How should such insurance be cor-rectly priced? Wars became more expensive. Under the old system, a subject's duties towards his monarch included military service. The new ways of mak-ing war required professionals and professionals had to be paid.[5] Governments started to borrow and bankers started to lend to them (a very high risk business,

---

[4] *Uncia*, from which we get the English *inch* and *ounce*. However, there are 16 ounces in an English pound and 20 fluid ounces in an English pint (but 16, slightly different, fluid ounces in a US pint). The South Australian pint contained 15 fluid ounces.

[5] 'Pas d'argent, pas de Suisse.' If you do not pay them, your Swiss mercenaries will not fight.

but very profitable if you got it right). Loans taken out over a long period required the computation of compound interest. All these things required fractions.

But how and why do these fractions and fractional calculations work? If the matter is one of intuition, why did the great ancient civilisations fail to develop such an 'intuitive' concept? If the answer is that transactions based on fractions always work, why did the Spanish Crown, with access to all the wealth of the Americas, declare bankruptcy four times in half a century?

For the remainder of the chapter, instead of hand-waving, we shall assume that any satisfactory answer must require work and thought. We shall try to construct the strictly positive rationals $\mathbb{Q}^+$ (that is to say, the positive fractions) from the natural numbers with the same care with which a watch maker might assemble a watch.

Exercise 2.1.5 *Von Neumann is quoted[6] as saying 'If people do not believe that mathematics is simple, it is only because they do not know how complicated life is.' When judging the difficulty of the rest of this book it may be useful to think about the following tasks of varying difficulty.*

(i) Everybody can walk. Write down instructions for walking in sufficient detail to enable an engineer to produce a two-legged walking robot.

(ii) Everybody can talk. Write down instructions on how to construct and recognise a sentence in your own language.

(iii) Choose a language that you do not speak. Take down a dictionary and a grammar book. Translate the first page of a book in the other language into your own. Now translate the first page of a book in your language into the other language. Which task is harder and why?

(iv) Choose a game of patience. Explain the rules to someone who has never seen a pack of cards.

(v) Choose a game of patience. Explain the rules to someone who is familiar with the rules of other patience games.

2.2 Equivalence Classes

Our first problem is to find a suitable language to talk about the positive rationals. Clearly, saying 'a positive rational is a fraction' is to define one unknown object in terms of another. We could try to improve on this by saying 'a positive rational is an object written p/q with p and q natural numbers', but we

---

[6] By Franz L. Alt in *Communications of the ACM*, volume 15, issue 7, July 1972, page 694.

then have to explain that objects like 2/3 and 4/6 'are really the same' when they are obviously different. A better approach might be to say that 2/3, 4/6, ... 'represent the same abstract object' but what is that 'abstract object'?

Mathematicians have invented the idea of an *equivalence class* to perform the role of 'abstract object' in this and similar contexts. If the reader feels that we are constructing a sledge hammer to crack a nut, she should reflect that some nuts are hard to crack and that, once we have our sledge hammer, we can crack lots of different nuts in the same way. We will use the notion of equivalence class in many of the constructions in this book and it occurs in the 'first chapter' of several subjects of pure mathematics (quotient groups in algebra, quotient topologies in analysis, atlases in differential geometry and so on).

We say that a collection A of mathematical objects is a *set*[7] and write $a \in A$ to mean that a belongs to A and $a \notin A$ to mean that a does not. We say that a set B is subset of A if, whenever $b \in B$, we have $b \in A$ (in other words, A contains B). In this section, we will be interested in a relation $\sim$ which may or may not exist between elements of A. We write $a \sim b$ if a is related to b and $a \nsim b$ if not.

Here are three properties, which may or may not hold, for a relation $\sim$ on A.

(1) $a \sim a$ for all $a \in A$. (We say that $\sim$ is *reflexive*.)
(2) If $a \sim b$, then $b \sim a$. (We say that $\sim$ is *symmetric*.)
(3) If $a \sim b$ and $b \sim c$, then $a \sim c$. (We say that $\sim$ is *transitive*.)

Exercise 2.2.1 *Consider the relations* $=$, $\geq$ *and* $>$ *on* $\mathbb{N}^+$. *Which are reflexive, which are symmetric and which are transitive?*

Exercise 2.2.2 *Consider a set* $X = \{x, y, z\}$ *containing three distinct elements* x, y *and* z. *In each of the following eight examples we define a relation* $\sim$ *which holds in the specified cases and no other. State, with reasons, whether the relation is reflexive, whether it is symmetric and whether it is transitive.*

   (i) $x \sim y, y \sim z$.
   (ii) $x \sim x, y \sim y, z \sim z, x \sim y, y \sim z$.
   (iii) $x \sim y, y \sim x$.
   (iv) $x \sim y, y \sim z, x \sim z$.
   (v) $x \sim y, y \sim x, x \sim x, y \sim y$.

---

[7] Henceforward we shall use the word collection to mean a collection of mathematical objects, that is to say, a set.

(vi) $x \sim x$, $y \sim y$, $z \sim z$, $x \sim y$, $y \sim z$, $x \sim z$.
(vii) $x \sim x$, $y \sim y$, $z \sim z$, $x \sim y$, $y \sim x$, $y \sim z$, $z \sim y$.
(viii) $x \sim x$, $y \sim y$, $z \sim z$.

Conclude that relations can have any combination of our three properties.

Exercise 2.2.3 *Consider a set X with a relation $\sim$ which is symmetric and transitive. Suppose that, given any $x \in X$, we can find a $y \in X$ with $x \sim y$ (that is to say, everything is related to something). Show that $\sim$ is actually reflexive.*

We now make the following definitions.

Definition 2.2.4 *Consider a set A with a relation $\sim$. We say that $\sim$ is an* equivalence relation *if it is reflexive, symmetric and transitive.*

Definition 2.2.5 *Consider a set A with an equivalence relation $\sim$. If $a \in A$, we write $[a]$ for the set of all $b \in A$ such that $b \sim a$ and call $[a]$ the* equivalence class *of a. We denote the collection of all such equivalence classes by $A/\sim$.*

We make the following remarks.

Lemma 2.2.6 *Suppose that we are in the situation described in Definition 2.2.5. Then the following statements are true.*

(i) $a \in [a]$.
(ii) *If $a \in [b]$, then $[a] = [b]$.*

Proof. (1) By reflexivity, $a \sim a$ and so $a \in [a]$.

(ii) If $x \in [a]$, then $x \sim a$ and $a \sim b$ so, by transitivity, $x \sim b$. Thus every member of $[a]$ is a member of $[b]$.

On the other hand, if $y \in [b]$, we have $y \sim b$ and $a \sim b$. By symmetry, $b \sim a$ and so, by transitivity, $y \sim a$. Thus every member of $[b]$ is a member of $[a]$. ∎

How can we use this idea to construct the strictly positive rationals? We think of the notion of a fraction a/a' already familiar from school and consider the set A of all (a, a') with a and a' natural numbers. (Note that, as the reader will expect, $(a, a') = (b, b')$ if and only if $a = b$ and $a' = b'$. The technical term for such an (a, a') is an *ordered pair*. Throughout this book we shall only deal with *ordered* pairs.) We define a relation $\sim$ on A by the condition $(a, a') \sim (b, b')$ if and only if $a \times b' = a' \times b$. (Recall the school idea of 'multiplying out fractions' by replacing $a/a' = b/b'$ with $a \times b' = a' \times b$.)

Lemma 2.2.7 *The relation $\sim$ just defined is, indeed, an equivalence relation.*

Proof. (i) Since $a \times a' = a' \times a$, we have $(a, a') \sim (a, a')$. Thus $\sim$ is reflexive.

(ii) If $(a, a') \sim (b, b')$, then $a \times b' = a' \times b$, so $a' \times b = a \times b'$ and $(b, b') \sim (a, a')$. Thus $\sim$ is symmetric.

(iii) This is the only non-trivial part. Suppose that $(a, a') \sim (b, b')$ and $(b, b') \sim (c, c')$. Then $a \times b' = a' \times b$ and $b \times c' = c \times b'$. It follows, using the associative and commutative laws for multiplication (see page 17), that

$$(a \times c') \times b' = b' \times (a \times c') = (b' \times a) \times c'$$
$$= (a \times b') \times c' = (a' \times b) \times c'$$
$$= a' \times (b \times c') = a' \times (b' \times c)$$
$$= a' \times (c \times b') = (a' \times c) \times b'.$$

The cancellation law law for multiplication (see page 19) yields

$$a \times c' = a' \times c,$$

so $(a, a') \sim (c, c')$ and we are done.   ■

Exercise 2.2.8 *Check the use of the appropriate law at each stage of the long calculation in the proof of Lemma 2.2.7 (iii).*

In future, we will often not do calculations like the sequence in the first paragraph of the proof of Lemma 2.2.7 explicitly, but say, 'using the associative and commutative laws of multiplication (or addition)', leaving it to any suspicious reader to write out the steps.

Exercise 2.2.9 *Show, using the associative and commutative laws explicitly, that*

$$\big((a \times b) \times c\big) \times d = \big((d \times c) \times b\big) \times a.$$

We now define addition and multiplication on the collection $A/\sim$ of equivalence classes. We need to proceed with care, as the next exercise shows.

Exercise 2.2.10 *We might be tempted to write*

$$[(a, a')] \boxplus [(b, b')] \underset{?}{=} [(a + b, a' + b')].$$

Show that $[(1, 2)] = [(2, 4)]$ and $[(1, 1)] = [(1, 1)]$, but

$$[(1 + 1, 2 + 1)]) \neq \big[((2 + 1), (4 + 1))\big].$$

Conclude that our putative definition makes no sense.

If we define an operation on equivalence classes by considering representative elements, we must ensure that we get the same result *whichever representative we choose.*

Lemma 2.2.11 *If $(a, a') \sim (b, b')$ and $(n, n') \sim (m, m')$, then*

$$\big((a \times n') + (a' \times n), a' \times n'\big) \sim \big((b \times m') + (b' \times m), b' \times m'\big).$$

Thus we may define

$$[(a, a')] \oplus [(n, n')] = \Big[\big((a \times n') + (a' \times n), a' \times n'\big)\Big]$$

unambiguously.

Proof. If we can show that

$$\big((a \times n') + (a' \times n), a' \times n'\big) \sim \big((a \times m') + (a' \times m), a' \times m'\big) \quad \bigstar$$

and

$$\big((a \times m') + (a' \times m), a' \times m'\big) \sim \big((b \times m') + (b' \times m), b' \times m'\big), \quad \bigstar\bigstar$$

the required result will follow from the transitivity of $\sim$.

Let us prove $\bigstar$. Since $(n, n') \sim (m, m')$, we have $n' \times m = m' \times n$, justifying step (2) in the next calculation.

$$
\begin{aligned}
\big((a \times n') + (a' \times n)\big) \times (a' \times m') &= (a' \times m') \times \big((a \times n') + (a' \times n)\big) \\
&= \big((a' \times m') \times (a \times n')\big) + \big((a' \times m') \times (a' \times n)\big) \quad (1) \\
&= \big((a \times m') \times (a' \times n')\big) + \big((a' \times a') \times (m' \times n)\big) \\
&= \big((a \times m') \times (a' \times n')\big) + \big((a' \times a') \times (n' \times m)\big) \quad (2) \\
&= \big((a' \times n') \times (a \times m')\big) + \big((a' \times n') \times (a' \times m)\big) \\
&= (a' \times n') \times \big((a \times m') + (a' \times m)\big) \quad (3) \\
&= \big((a \times m') + (a' \times m)\big) \times (a' \times n').
\end{aligned}
$$

Steps (1) and (3) are justified by the distributive law (see page 17) and the unnumbered steps by repeated use of the associative and commutative laws of multiplication.

The formula $\bigstar\bigstar$ can be proved similarly or, more quickly, by using $\bigstar$ and the commutative law of multiplication. I leave the choice and the work to the reader. ∎

Exercise 2.2.12 *Prove $\bigstar\bigstar$.*

Remark In stating Lemma 2.2.11, we have been guided by the formula

$$\frac{a}{a'} + \frac{b}{b'} \underset{?}{=} \frac{ab' + a'b}{a'b'}.$$

However, we have not established this formula, or even shown that it makes sense. (Is it supposed to apply to 'fractions', whatever they may be, or to rational numbers each of which is 'represented' by many different fractions?) We break the vicious circle by *defining* the operation of addition of positive rationals before trying to *prove* results about that operation.

Exercise 2.2.13 *If* $(a, a') \sim (b, b')$ *and* $(n, n') \sim (m, m')$, *show that*

$$(a \times n, a' \times n') \sim (b \times m, b' \times m').$$

Conclude that we may define

$$[(a, a')] \otimes [(n, n')] = [(a \times n, a' \times n')]$$

unambiguously.

Which fractional 'equality' are we thinking of?

At this point the reader may object that it is unfair to leave parts of the development to her. I remind the reader that, as stated in the introduction, solutions to exercises like this one will be found on my home page. If this does not satisfy the reader, then, once again, I recommend Landau's *Foundations of Analysis* [20], where every step is carefully laid out.

Lemma 2.2.14 *If* $(a, a') \sim (b, b')$, $(n, n') \sim (m, m')$ *and* $a \times n' > a' \times n$, *then* $b \times m' > b' \times m$.

Thus we may define a relation $\ominus$ *on* $A/\sim$ *by taking* $[(a, a')] \ominus [(n, n')]$ *if and only if* $a \times n' > a' \times n$.

Proof. Just as in Lemma 2.2.11, we do the proof in two stages, first showing that $a \times m' > a' \times m$.

To this end, we observe that

$$
\begin{aligned}
(a \times m') \times n' &= m' \times (a \times n') \\
&> m' \times (a' \times n) \qquad &(1) \\
&= a' \times (n \times m') \\
&= a' \times (m \times n') \qquad &(2) \\
&= (a' \times m) \times n'.
\end{aligned}
$$

(Step (1) uses the multiplication law for inequalities from Lemma 1.3.8. Step (2) uses the fact that $(n, n') \sim (m, m')$. The remaining steps use the associative and commutative laws of multiplication, sometimes condensing several steps into one.) The cancellation law for multiplication (see Lemma 1.3.9) now gives

$$a \times m' > a' \times m,$$

as required.

A very similar argument now allows us to deduce that $b \times m' > b' \times m$. ∎

Exercise 2.2.15 *Give the similar argument just invoked.*

We note a useful cancellation law.

Lemma 2.2.16 *We have* $[(c \times a, c \times a')] = [(a, a')]$.

Proof. This just uses the associative and commutative laws of multiplication to obtain

$$(c \times a) \times a' = c \times (a \times a') = c \times (a' \times a) = (c \times a') \times a. \qquad ∎$$

2.3 Properties of the Strictly Positive Rationals

We have defined a collection of equivalence classes $A/\sim$ with associated operations $\oplus$ and $\otimes$ and a relation $\ominus$. We now need to show that they have the properties we would expect of the strictly positive rationals. To do this we need to state *explicitly* what these properties are, and this accounts for the list-like character of the next theorem. The hardest work has already been done when we produced our definitions and showed that they were coherent and, though the theorem is composed of many sub-theorems, each of them is easy to prove.

From now on, we economise on brackets by writing $[a, a'] = [(a, a')]$. Since we shall use bold-face letters like **a** frequently, the reader may find it useful to recall that these are usually hand written as $\underline{a}$ or $\underline{a}$.

Theorem 2.3.1 *Let us write general elements of $A/\sim$ as* **a**, **b**, ... *Then the following results hold.*

(i) $\mathbf{a} \oplus \mathbf{b} = \mathbf{b} \oplus \mathbf{a}$. *(Commutative law of addition)*
(ii) $\mathbf{a} \oplus (\mathbf{b} \oplus \mathbf{c}) = (\mathbf{a} \oplus \mathbf{b}) \oplus \mathbf{c}$. *(Associative law of addition)*
(iii) $\mathbf{a} \otimes \mathbf{b} = \mathbf{b} \otimes \mathbf{a}$. *(Commutative law of multiplication)*
(iv) $\mathbf{a} \otimes (\mathbf{b} \otimes \mathbf{c}) = (\mathbf{a} \otimes \mathbf{b}) \otimes \mathbf{c}$. *(Associative law of multiplication)*

(v) If we write $\mathbf{1} = [1, 1]$*, then* $\mathbf{1} \times \mathbf{a} = \mathbf{a}$*. (Existence of a multiplicative unit)*

(vi) For each $\mathbf{a}$ *we can find an* $\mathbf{a}^{-1}$ *such that* $\mathbf{a} \times \mathbf{a}^{-1} = \mathbf{1}$*. (Existence of a multiplicative inverse)*

(vii) $\mathbf{a} \otimes (\mathbf{b} \oplus \mathbf{c}) = (\mathbf{a} \otimes \mathbf{b}) \oplus (\mathbf{a} \otimes \mathbf{c})$*. (Distributive law)*

(viii) We can find a $\mathbf{c}$ *such that* $\mathbf{a} = \mathbf{b} \oplus \mathbf{c}$ *if and only if* $\mathbf{a} \ominus \mathbf{b}$*. (Order rule)*

(ix) Exactly one of the following conditions holds: $\mathbf{a} \ominus \mathbf{b}$ *or* $\mathbf{b} \ominus \mathbf{a}$ *or* $\mathbf{a} = \mathbf{b}$*. (Trichotomy)*

Proof. As I said above, all of these results are now easy to demonstrate. (The reader may murmur under her breath 'After all, we are just dealing with fractions in a light disguise.') I shall prove a selection, leaving the rest as an exercise. We take $\mathbf{a} = [a, a']$, $\mathbf{b} = [b, b']$ and so on.

(v) Because 1 is a unit for multiplication of natural numbers (see page 17),

$$\mathbf{1} \otimes \mathbf{a} = [1, 1] \otimes [a, a'] = [1 \times a, 1 \times a'] = [a, a'] = \mathbf{a}.$$

(vi) Observe that, if $(a, a') \sim (b, b')$, then $a \times b' = a' \times b$ so $a' \times b = a \times b'$ and $(a', a) \sim (b', b)$. Thus we may define $\mathbf{a}^{-1} = [a', a]$. The commutative law of multiplication for the natural numbers now tells us that

$$\mathbf{a} \times \mathbf{a}^{-1} = [a, a'] \otimes [a', a] = [a \times a', a' \times a] = [a \times a', a \times a'] = [1, 1] = \mathbf{1}.$$

(vii) We have

$$
\begin{aligned}
\mathbf{a} \otimes (\mathbf{b} \oplus \mathbf{c}) &= [a, a'] \otimes \big[(b \times c') + (c \times b'), b' \times c'\big] \\
&= \big[a \times ((b \times c') + (c \times b')), a' \times (b' \times c')\big] \\
&= \big[(a \times (b \times c')) + (a \times (c \times b')), a' \times (b' \times c')\big] \quad (1) \\
&= \big[a \times (b \times c'), a' \times (b' \times c')\big] \oplus \big[a \times (c \times b'), a' \times (b' \times c')\big] \\
&= \big[c' \times (a \times b), c' \times (a' \times b')\big] \oplus \big[b' \times (a \times c), b' \times (a' \times c')\big] \\
&= [a \times b, a' \times b'] \oplus [a \times c, a' \times c'] \quad (2) \\
&= (\mathbf{a} \otimes \mathbf{b}) \oplus (\mathbf{a} \otimes \mathbf{c}).
\end{aligned}
$$

For step (1) we used the distributive law for natural numbers and for step (2) the cancellation law stated as Lemma 2.2.16. The remaining steps use definitions and the associative and commutative laws of multiplication for the natural numbers.

(viii) If $\mathbf{a} \ominus \mathbf{b}$, then $a \times b' > a' \times b$, so, by the order rule for the natural numbers, we can find a d such that $a \times b' = (a' \times b) + d$. If we set $\mathbf{c} = [d, a' \times b']$, then, using the distributive law for the natural numbers together with the associative and commutative laws of multiplication, we have

$$\mathbf{b} \oplus \mathbf{c} = \left[(b \times (a' \times b')) + (b' \times d), b' \times (a' \times b') \right]$$
$$= \left[(b' \times (a' \times b)) + (b' \times d), b' \times (a' \times b') \right]$$
$$= \left[b' \times ((a' \times b) + d), b' \times (a' \times b') \right]$$
$$= \left[b' \times (a \times b'), b' \times (a' \times b') \right]$$
$$= [(b' \times b') \times a, (b' \times b') \times a'] = [a, a'] = \mathbf{a}.$$

Conversely, if $\mathbf{a} = \mathbf{b} \oplus \mathbf{c}$, then

$$(a, a') = ((b \times c') + (b' \times c), b' \times c'),$$

so, since $(b \times c') + (b' \times c) > b \times c'$, we have, using the multiplication law for inequalities given in Lemma 1.3.8 (ii),

$$(a \times b') \times c' = a \times (b' \times c') = a' \times ((b \times c') + (b' \times c))$$
$$> a' \times (b \times c') = (a' \times b) \times c'.$$

The cancellation law for multiplication in $\mathbb{N}^+$ (Lemma 1.3.9 (ii)) now gives $a \times b' > a' \times b$ so $\mathbf{a} \ominus \mathbf{b}$ and we are done. ∎

Exercise 2.3.2 *Prove the remaining parts of Theorem 2.3.1.*

Exercise 2.3.3 *Show that any system satisfying the conditions laid out in Theorem 2.3.1 will also obey the following rules.*

(ix) If $\mathbf{a} \ominus \mathbf{b}$ and $\mathbf{b} \ominus \mathbf{c}$, then $\mathbf{a} \ominus \mathbf{c}$. (Transitivity of order)
(x) If $\mathbf{a} \ominus \mathbf{b}$, then $\mathbf{a} \oplus \mathbf{c} \ominus \mathbf{b} \oplus \mathbf{c}$. (Order and addition)
(xi) If $\mathbf{a} \ominus \mathbf{b}$, then $\mathbf{a} \otimes \mathbf{c} \ominus \mathbf{b} \otimes \mathbf{c}$. (Order and multiplication)

Prove the cancellation laws

(a) If $\mathbf{a} \oplus \mathbf{c}, \ominus \mathbf{b} \oplus \mathbf{c}$, then $\mathbf{a} \ominus \mathbf{b}$.
(b) If $\mathbf{a} \otimes \mathbf{c} \ominus \mathbf{b} \otimes \mathbf{c}$ then $\mathbf{a} \ominus \mathbf{b}$.

[See Lemmas 1.3.4, 1.3.8 and 1.3.9 if you need a hint, but notice that the proof of (b) is simple using the existence of a multiplicative inverse.]

With one exception, the laws discussed above apply to the natural numbers. The novelty is the existence of a *multiplicative inverse* $\mathbf{a}^{-1}$.

We have shown the existence of a multiplicative unit and a multiplicative inverse as Theorem 2.3.1 (v) and (vi). We now need to check that they are unique.

Lemma 2.3.4 *Consider any system satisfying the conclusions of Theorem 2.3.1.*

(i) If $\tilde{1} \otimes a = a$ for all a, then $\tilde{1} = 1$.

(ii) If $a \otimes a^* = 1$, then $a^* = a^{-1}$.

Proof. (i) Observe that, taking $a = 1$, we have

$$1 = \tilde{1} \otimes 1 = 1 \otimes \tilde{1} = \tilde{1},$$

where we justify the calculation by using the commutativity of $\otimes$ and the fact that 1 is a multiplicative unit (see Theorem 2.3.1 (iii) and (v)).

(ii) Using the commutative and associative laws for $\otimes$, we have

$$a^* = 1 \otimes a^* = (a \otimes a^{-1}) \otimes a^* = (a^{-1} \otimes a) \otimes a^*$$
$$= a^{-1} \otimes (a \otimes a^*) = a^{-1} \otimes 1 = 1 \otimes a^{-1} = a^{-1},$$

as required. ∎

We obtain a couple of useful corollaries.

Lemma 2.3.5 *Consider any system satisfying the conclusions of Theorem 2.3.1.*

(i) We have $(a^{-1})^{-1} = a$.

(ii) We have $(a \otimes b)^{-1} = a^{-1} \otimes b^{-1}$.

Proof. (i) Using the commutative law of multiplication,

$$a^{-1} \otimes a = a \otimes a^{-1} = 1;$$

so, by the uniqueness of the multiplicative inverse, $(a^{-1})^{-1} = a$.

(ii) Using the commutative and associative laws of multiplication,

$$(a \otimes b) \otimes (a^{-1} \otimes b^{-1}) = (a \otimes a^{-1}) \otimes (b \otimes b^{-1}) = 1 \otimes 1 = 1;$$

so, by the uniqueness of the multiplicative inverse, $(a \otimes b)^{-1} = a^{-1} \otimes b^{-1}$. ∎

If we think about the relation between fractions and 'whole numbers' in school arithmetic, it becomes clear that we would like to say that 'the strictly positive rationals contain a copy of the natural numbers'. The difficulty, if any, in showing this lies, not in the proof, but in finding a precise way to say what we mean. To this end, we introduce a simple definition.

Definition 2.3.6 *We say that a function $f : X \to Y$ is* injective *if $f(x) = f(x')$ implies $x = x'$ (that is to say, f carries different elements of X to different members of Y).*

Note We write $f : X \to Y$ to mean a function taking members of X to members of Y. I think of a function $f : X \to Y$ as a machine which, given a particular $x \in X$, produces a unique element of Y denoted by $f(x)$.

We can now state what we wish to prove.

Lemma 2.3.7 *There is an injective function $f : \mathbb{N}^+ \to A/\sim$ such that*

$$f(n + m) = f(n) \oplus f(m), \quad f(n \times m) = f(n) \otimes f(m)$$

$$and \ n > m \ implies \ f(n) \ominus f(m).$$

Proof. Not surprisingly, we set $f(n) = [n, 1]$. If $f(n) = f(m)$, then $(n, 1) \sim (m, 1)$, and so, by definition, $n = n \times 1 = m \times 1 = m$. Thus f is injective. The remaining verifications are equally easy.

$$f(n) \oplus f(m) = [n, 1] \oplus [m, 1] = [(n \times 1) + (m \times 1), 1 \times 1] = [n + m, 1]$$
$$= f(n + m)$$
$$f(n) \otimes f(m) = [n, 1] \otimes [m, 1] = [n \times m, 1 \times 1] = [n \times m, 1] = f(n \times m).$$

Finally, if $n > m$, then $n \times 1 = n > m = m \times 1$ and so

$$f(n) = [n, 1] \ominus [m, 1] = f(m). \qquad \blacksquare$$

Lemma 2.3.7 allows us to use 2 as an abreviation for $[2, 1]$.

2.4 What Have We Actually Done?

Goethe said that 'Mathematicians are a sort of Frenchmen; if you talk to them, they translate it into their own language, and then it is immediately something quite different.' The reader can legitimately complain that the previous section just takes the obvious and makes it as incomprehensible as possible. For example, the first part of proof of Theorem 2.3.1 (viii) is 'just the calculation'

$$\frac{b}{b'} + \frac{ab' - a'b}{a'b'} = \frac{ba'b' + b'ab' - a'bb'}{a'b'b'} = \frac{b'b'a}{b'b'a'} = \frac{a}{a'}.$$

In reply, I would ask the reader how much she remembers of learning about fractions in school. Presumably she learned how to calculate with fractions and about the way fractions behave. Does the reader remember being shown that the rules of behaviour were consistent with the methods of calculation?[8] Most of my readers obtained high marks in examinations involving fractions, but this

---

[8] Notice that I do not say that your teacher failed to demonstrate consistency. I merely doubt that you remember whether she did or not.

only shows that they do well in situations where agreeing with the examiner produces high marks.

It could be argued that we have clear intuitive understanding of fractions and this is all we need. Thus, we intuitively understand that

$$\frac{3}{14} + \frac{1}{10} = \frac{11}{35}.$$

Pressed to explain where this intuitive feeling comes from, we could explain that we take one apple pie, cut it into 14 equal parts and set aside 3 of them. We now take another apple pie, cut it into 10 equal parts and set aside 1 of them. We then cut each of the set aside slices from the first apple pie into 10 equal parts, obtaining the equivalent of 30 slices of an apple pie cut into 140 parts and cut the set aside slice from the second apple pie into 14 equal parts, obtaining the equivalent of 14 slices of an apple pie cut into 140 parts. Since the set aside slices from the two apple tarts are now of equal size, we can put them together to obtain the equivalent of 44 slices of an apple pie cut into 140 parts. We now recombine the slices 4 at a time to obtain the equivalent of 11 slices from an apple pie cut into 35 equal bits.

Asked what we do about the crumbs and general stickiness, we reply that we are not thinking about real-life apple pies, but idealised apple pies which come as identical[9] disks which can be divided into as many equal sectors[10] as we wish. If we are going to operate with this kind of idealised object, surely it is better to stick with the idealised objects $\mathbb{N}^+$ and $\mathbb{Q}^+$ rather than introduce a new sort of idealised objects including an inexhaustible source of 'perfect' apple pies and a 'perfectly sharp' knife.

If we are sure that fractions work, but we are not sure why we are sure why they work, then it is reasonable to take the rules for calculation that we were taught in school and check that the results are indeed consistent with the behaviour we expect from fractions. If we adopt a different notation from that which we normally use, then we will have less chance of skipping over essential steps. This is precisely what we did in the previous section.

Of course, this procedure moves fractions away from the real world of money changers and apple pie sellers. However, we are free to decide whether and in what respect our abstract rational numbers apply to the real world. Farmers may decide that half a sheep is not a useful idea, while butchers may be perfectly happy with the concept. Modern physics does not deal with one twelfth of an electron or half a photon but is happy with half a metre.

---

[9] Whatever that means. See Section 3.4 for further discussion.
[10] A sector is 'the portion of a disk enclosed by two radii and an arc'.

3

The Rational Numbers

3.1 Negative Numbers

We might expect that, as people got more used to the Indian system, they would move from treating ∗ as a placeholder to treating it 'as if it were a number' and finally saying that ∗, or 0, as we shall now call it, was a number just like any other. We would not expect the process to be easy. If, after saying that a field contains three sheep and two cows, you remark that it contains zero horses, should you not add that it contains zero swans and zero hippopotamuses?

> 'I see nobody on the road,' said Alice.
> 'I only wish I had such eyes,' the King remarked in a fretful tone. 'To be able to see Nobody! And at that distance, too! Why, it's as much as I can do to see real people, by this light!'
>
> *Lewis Carroll*, Alice through the Looking-Glass

It must also be said that the admonition to treat zero like any other number loses some of its force when coupled with a further admonition *never* to try and divide by zero.

It is thus not surprising that it took a thousand years for everybody to accept zero as a number. What is surprising is that, at the beginning of those thousand years, the great Indian mathematician Brahmagupta not only considered zero as a number, but also went on to introduce entirely new objects which we now call 'negative numbers' and give what we now consider the correct rules[1] for operating with them.

If, as Brahmagupta suggested, we consider negative numbers as debts, so that a fortune of -3 gold pieces is actually a debt of 3 gold pieces, then adding

---

[1] His rule for $0/0$ is a minor exception. Since we cannot extend *both* of the desirable formulae $1 = a/a$ and $0/a = 0$ to the case $a = 0$ simultaneously, we leave $0/0$ *undefined*.

39

7 to −5 is like paying off a debt of 5 gold pieces out of a fortune of 7 gold pieces, leaving a fortune of 2 gold pieces. Thus

$$7 + (-5) = 2.$$

On the other hand, if we add 4 to −6, this is like trying to pay off a debt of 6 gold pieces with a fortune of 4 gold pieces. We will remain 2 gold pieces in debt, so

$$4 + (-6) = -2.$$

So far so good (and similar ideas can be found a few centuries earlier in China), but what rules should we use for multiplication?

Before answering the question, we should perhaps ask a more fundamental question. Who cares? If we look at the farmer, the merchant, the tax collector and even the astronomer, the answer must be none of them. Calling a debt of 5 gold pieces an asset of −5 gold pieces makes it no easier to pay.

We need to introduce the mathematical puzzler, the kind of person who delights in finding the initial number of walnuts in this question from Metrodorus, who made a collection of such problems in the sixth century AD.

Exercise 3.1.1 *Mother, why dost thou pursue me with blows on account of the walnuts? Pretty girls divided them all among themselves. For Melission took two-sevenths of them from me, and Titane took the twelfth. Playful Astyoche and Philinna have the sixth and third. Thetis seized and carried off twenty, and Thisbe twelve, and look there at Glauce smiling sweetly with eleven in her hand. This one nut is all that is left to me.*

We can call such problems 'puzzles' because they have no practical use. Some of these problems would now be stated in terms of a linear equation in one variable, some as simultaneous equations in two variables and some as quadratic equations. Historians have traced these kinds of problems back to the ancient Babylonians. Standing as we do on the shoulders of giants, we find them easy to classify and then solve in a uniform manner.

Here are some quadratic problems from the work of Bhāskarāchārya, perhaps the greatest mathematician of the classical period of Indian mathematics.

Exercise 3.1.2 *The eighth part of a troop of monkeys, squared was skipping in a grove and delighted with their sport. Twelve remaining were seen on the hill, amused with chattering to each other. How many were there? (All translations are taken from [5].)*

[*In modern notation,* $(n/8)^2 + 12 = n$, *where n is the number of monkeys.*]

Exercise 3.1.2 has two possible solutions.

Exercise 3.1.3

(i) *Construct a 'monkey puzzle' with one strictly positive integer solution and one strictly negative integer solution. (So the puzzle has exactly one 'appropriate' solution.)*

(ii) *Construct a 'monkey puzzle' with two strictly negative integer solutions. (So the puzzle has no 'appropriate' solution.)*

The solver of 'monkey puzzles' has three possible strategies.

(1) Set out the instructions for solving such a puzzle in such a way that negative numbers never appear. Al-Khwārizmī gave a collection of such instructions, but this required splitting the problem into a number of cases. In modern algebraic notation, a quadratic with non-zero coefficients and leading coefficient 1 might need to be written

$$x^2 + bx + c = 0, \ x^2 + bx = c, \ x^2 + c = bx \text{ or } x^2 = bx + c$$

to ensure that b and c are positive.

(2) Work through the problem 'as if negative numbers exist and behave in the right way'. At the end reject all those solutions which are inappropriate, including any 'negative numbers'.

(3) State firmly that negative numbers exist and behave in the right way. Negative solutions of a quadratic are solutions just like any other. If the problem is about gold coins, the negative solutions represent debts. If the problem is about monkeys, then the problem is one about finding all positive integer solutions to a quadratic.

If we look at European mathematicians between 1300 and 1600, we can find examples of each of these strategies, and, no doubt, some people used different strategies at different times or on different occasions,

There is a further problem with the first two strategies, illustrated by another monkey problem from Bhāskarāchārya.

Exercise 3.1.4 *The square of the fifth part of a group of monkeys less three [that is to say, $(n/5 - 3)^2$, where n is the number of monkeys] had gone into a cave; and one monkey was in sight, having climbed on a branch. Say how many there were.*

At first sight, this has the two reasonable solutions, 50 and 5, but, if $n = 5$, then $n/5 - 3$ is a negative integer and, says Bhāskarāchārya, this solution is not to be taken 'For it is incongruous. People do not approve a negative absolute number.'

Here is another problem, where I have modified the original numbers to make a point.

Exercise 3.1.5 *Four pairs out of a flock of geese remained sporting in the water and saw seven times the square root of the flock proceeding to the shore tired of the diversion. Tell me dear girl[2] the number of the flock. [In modern notation, $8 + 7\sqrt{n} = n$, where n is the number of geese and we take the positive square root.]*

Check your answer by substituting back in the problem. Why is there only one appropriate answer?

As mathematicians began to use symbolic algebra (so that letters replaced first unknowns and, later, variable parameters), it became harder to ensure that negative numbers never entered calculations in some disguise or another. However, the most satisfactory methods for proving things remained geometrical, and the 'natural' way of viewing geometrical quantities allows only positive numbers.[3]

As we shall see with the complex numbers and vectors, new systems are often adopted not because they are easier to use for existing purposes, but because they lend themselves to new uses. In the end, objections to negative numbers were overwhelmed by the new uses which resulted from the scientific revolution.

Ordinary people used the new thermometer whose scale contained zero and negative numbers. Today northern city dwellers are more likely to say 'It is minus five today, so dress up very warm' than 'There are five sheep in the field.' War histories talk about 'D-day minus two' or 'zero hour'.[4] Descartes's invention of coordinate geometry leads directly to the graphs which adorn our serious magazines and which freely use both positive and negative numbers ('the stock market continued in negative territory for the third day running').

Oceanic voyages required new astronomical tables and endless calculations to prepare and interpret them. The invention of decimals and logarithms (see

---

[2] Bhāskarāchārya is supposed to have written his book as a present for his daughter.

[3] Modern treatments do allow 'signed quantities' so that angles, lengths and areas can be negative.

[4] Notice the contrast with the older calendar convention of going directly from 1 BC to 1 AD.

page 145) allowed addition to be substituted for multiplication, but required the use of negative numbers.[5]

The citizen of today's world thus has an 'intuitive idea' of negative numbers and the laws which govern their order and addition. I doubt if most citizens have an intuitive idea of the laws which govern the multiplication of negative numbers.

Minus times minus results in a plus,
The reasons for this we need not discuss.[6]

I have seen several different 'intuitive arguments' for the 'minus times minus' rule and accept that the authors of these arguments find them 'intuitively convincing'. However, evidence that the average citizen finds these various arguments intuitively convincing seems to be lacking.

Surely, Brahmagupta gave the rules he did not because he found them intuitive, but because they worked. I strongly urge the reader to experiment as to which multiplication rules allow the extended system of positive and negative numbers to behave as closely as possible to the system of strictly positive numbers. She will find that only the standard rules 'minus times plus results in minus' and 'minus times minus results in plus' will allow rules like the distributive law

$$a \times (b + c) = (a \times b) + (a \times c)$$

to be extended to the new system. (See Exercises 3.2.11 and 3.2.14, if you want to make this precise.)

However, the fact that Brahmagupta's rules are the only ones that could *possibly* work does not prove that they *actually* work. We thus have, as usual, three choices.

(1) To shout very loudly that they work. Words like 'common sense' and 'intuition' are very useful in this context.
(2) To say that generation after generation has used Brahmagupta's rules in all sorts of contexts and nothing has gone wrong. This 'engineer's justification' seems to me a very good one.[7]

---

[5] Recall that $\log(1/2) = -\log 2$. I am old enough to have spent many school hours calculating with logarithms. The representation actually used was $\log_{10} x = n + \log_{10} y$, where $10 > y \geq 1$ and n was an integer which could be positive or negative.

[6] A school mnemonic remembered without affection in Auden's *A Certain World*.

[7] As Heaviside complained 'Shall I refuse my dinner because I do not fully understand the process of digestion' (in his *Electromagnetic Theory*, Volume II, page 13). He would undoubtedly have classed this book with 'the long and disagreeable demonstrations [of the pure mathematician] to prove what [the physicist] *knows*'.

(3) To construct an extended system and check that it works.

This book being what it is, I shall pursue the third option.

3.2 Defining the Rational Numbers

We start from the strictly positive rational numbers $\mathbb{Q}^+$, equipped with the operations addition $+$, multiplication $\times$ and the relation greater than $>$ having the properties set out in the conclusions of Theorem 2.3.1 and Lemma 2.3.7.

In Section 2.2 we constructed the strictly positive rationals from the natural numbers (that is to say, the strictly positive integers) by using equivalence classes. If a trick works once, it is natural to try it again. Our first step is almost a word-for-word repetition.

We consider the set B of all (a, a') with a and a' in $\mathbb{Q}^+$. [The reader will recall from page 29 that $(a, a') = (b, b')$ if and only if $a = b$ and $a' = b'$.] If we think of a as our fortune and a' as our debt it is natural to define a relation $\sim$ on B by the condition $(a, a') \sim (b, b')$ if and only if $a + b' = a' + b$. (Notice that we cannot transform the last equation into the statement $a - a' \underset{?}{=} b - b'$, because the entire object of our discussion is to provide a framework in which such a statement might make sense.)

Exercise 3.2.1 *Show that the relation $\sim$ just defined is indeed an equivalence relation.*

We now define addition and multiplication on the collection $B/\sim$ of equivalence classes. Addition is easy.

Exercise 3.2.2 *If $(a, a') \sim (b, b')$ and $(c, c') \sim (d, d')$, show that*

$$(a + c, a' + c') \sim (b + d, b' + d').$$

Thus we may define

$$[(a, a')] \oplus [(b, b')] = [(a + b, a' + b')].$$

Multiplication, as one might expect, is more complicated, but it is not difficult (perhaps after a little experimentation) to see what is required.[8]

---

[8] Although we are not allowed to *assume* the formula

$$(a - a') \times (b - b') \underset{?}{=} ((a \times b) + (a' \times b')) - ((a \times b') + (a' \times b))$$

Lemma 3.2.3 *If $(a, a') \sim (b, b')$ and $(c, c') \sim (d, d')$, then*

$$\big((a \times c) + (a' \times c'), (a \times c') + (a' \times c)\big) \sim \big((b \times d) + (b' \times d'), (b \times d') + (b' \times d)\big).$$

Thus we may define

$$[(a, a')] \otimes [(b, b')] = \big[\big((a \times b) + (a' \times b'), (a \times b') + (a' \times b)\big)\big].$$

Proof. We first show that

$$\big((a \times c) + (a' \times c'), (a \times c') + (a' \times c)\big) \sim \big((a \times d) + (a' \times d'), (a \times d') + (a' \times d)\big).$$

To this end, observe that $c + d' = c' + d$, so

$$
\begin{aligned}
\big((a \times c) &+ (a' \times c')\big) + \big((a \times d') + (a' \times d)\big) \\
&= \big((a \times c) + (a \times d')\big) + \big((a' \times c') + (a' \times d)\big) \\
&= \big(a \times (c + d')\big) + \big(a' \times (c' + d)\big) \qquad\qquad (1) \\
&= \big(a \times (c' + d)\big) + \big(a' \times (c + d')\big) \\
&= \big((a \times c') + (a \times d)\big) + \big((a' \times c) + (a' \times d')\big) \qquad (2) \\
&= \big((a \times d) + (a' \times d')\big) + \big((a \times c') + (a' \times c)\big)
\end{aligned}
$$

as required. [In addition to the commutative and associative laws of addition and multiplication, we used the distributive law for $\mathbb{Q}^+$ at steps (1) and (2).]

A similar calculation (or an appropriate use of the commutative law of multiplication) now shows that

$$\big((a \times d) + (a' \times d'), (a \times d') + (a' \times d)\big) \sim \big((b \times d) + (b' \times d'), (b \times d') + (b' \times d)\big)$$

and the full result follows by the transitivity of $\sim$. ∎

Exercise 3.2.4 *Obtain the result stated in the last paragraph of the previous proof.*

Exercise 3.2.5 *Show that we may define a relation $\ominus$ on B by taking*

$$[(a, a')] \ominus [(b, b')]$$

if and only if $a + b' > a' + b$.
[Hint: You may find Lemma 1.3.9 (iii) useful.]

From now on, we economise on brackets by writing $[a, a'] = [(a, a')]$.

or even to claim that that the formula makes sense, we are allowed to use it as a source of ideas. The recommendation 'Drive as though everybody else is drunk' may be good advice even if based on questionable premises.

Exercise 3.2.6

(i) Show that $[a, a] = [1, 1]$ *and* $[a + 1, a] = [1 + 1, 1]$ *for all* $a \in \mathbb{Q}^+$.
(ii) Show that $[a, a'] \otimes [b, b'] = [a', a] \otimes [b', b]$.

Theorem 3.2.7 *Take* $\mathbf{0} = [1, 1]$, $\mathbf{1} = [1 + 1, 1]$ *and write general elements of* $B/\sim$ *as* $\mathbf{a}$, $\mathbf{b}$, $\mathbf{c}$, $\ldots$ *. Then the following results hold.*

(i) $\mathbf{a} \oplus \mathbf{b} = \mathbf{b} \oplus \mathbf{a}$. *(Commutative law of addition)*
(ii) $\mathbf{a} \oplus (\mathbf{b} \oplus \mathbf{c}) = (\mathbf{a} \oplus \mathbf{b}) \oplus \mathbf{c}$. *(Associative law of addition)*
(iii) $\mathbf{0} \oplus \mathbf{a} = \mathbf{a}$. *(Additive zero)*
(iv) For each $\mathbf{a}$ *we can find* $-\mathbf{a}$ *such that* $\mathbf{a} \oplus (-\mathbf{a}) = \mathbf{0}$. *(Additive inverse)*
(v) $\mathbf{a} \otimes \mathbf{b} = \mathbf{b} \otimes \mathbf{a}$. *(Commutative law of multiplication)*
(vi) $\mathbf{a} \otimes (\mathbf{b} \otimes \mathbf{c}) = (\mathbf{a} \otimes \mathbf{b}) \otimes \mathbf{c}$. *(Associative law of multiplication)*
(vii) $\mathbf{1} \otimes \mathbf{a} = \mathbf{a}$. *(Multiplicative unit)*
(viii) If $\mathbf{a} \neq \mathbf{0}$, *there exists an* $\mathbf{a}^{-1}$ *such that* $\mathbf{a} \otimes \mathbf{a}^{-1} = \mathbf{1}$. *(Existence of a multiplicative inverse)*
(ix) $\mathbf{a} \otimes (\mathbf{b} \oplus \mathbf{c}) = (\mathbf{a} \otimes \mathbf{b}) \oplus (\mathbf{a} \otimes \mathbf{c})$. *(Distributive law)*
(x) If $\mathbf{a} \ominus \mathbf{b}$ *and* $\mathbf{b} \ominus \mathbf{c}$, *then* $\mathbf{a} \ominus \mathbf{c}$. *(Transitivity of order)*
(xi) Exactly one of the following conditions holds: $\mathbf{a} \ominus \mathbf{b}$ *or* $\mathbf{b} \ominus \mathbf{a}$ *or* $\mathbf{a} = \mathbf{b}$. *(Trichotomy)*
(xii) If $\mathbf{a} \ominus \mathbf{b}$ *then* $\mathbf{a} \oplus \mathbf{c} \ominus \mathbf{b} \oplus \mathbf{c}$. *(Order and addition)*
(xiii) If $\mathbf{a} \ominus \mathbf{b}$ *and* $\mathbf{c} \ominus \mathbf{0}$ *then* $\mathbf{a} \otimes \mathbf{c} \ominus \mathbf{b} \otimes \mathbf{c}$. *(Order and multiplication)*
(xiv) $\mathbf{0} \neq \mathbf{1}$. *(Non-triviality)*

Proof. We prove a selection, leaving the remainder to the reader.

(iv) Observe that, if $(a, a') \sim (b, b')$, then $a + b' = a' + b$ so that $(a', a) \sim (b', b)$. Thus we may define $-[a, a'] = [a', a]$. We have

$$\mathbf{a} \oplus (-\mathbf{a}) = [a + a', a' + a] = [a + a', a + a'] = [1, 1] = \mathbf{0},$$

as required.

(viii) This is the least smooth part of the proof. Let $\mathbf{a} = [a, a']$. Since $\mathbf{a} \neq \mathbf{0}$, we have $a \neq a'$, so either $a > a'$ or $a' > a$ (trichotomy).

If $a > a'$, the order rule tells us that there exists a natural number c with $a = a' + c$. We observe that, if $(a, a') \sim (b, b')$, then

$$a + (b' + c) = (a + b') + c = (a' + b) + c = (a' + c) + b = a + b;$$

so, by the cancellation law for addition (Lemma 1.3.9 (i)), $b = b' + c$. Thus we may define $\mathbf{a}^{-1} = [1 + c^{-1}, 1]$ unambiguously.

Since $(a, a') = (a' + c, a') \sim (1 + c, 1)$ we have $\mathbf{a} = [1 + c, 1]$ and

$$
\begin{aligned}
\mathbf{a} \otimes \mathbf{a}^{-1} &= [1 + c, 1] \otimes [1 + c^{-1}, 1] \\
&= \left[\left((1 + c) \times (1 + c^{-1}) \right) + (1 \times 1), \left((1 + c) \times 1 \right) \right. \\
&\quad \left. + \left(1 \times (1 + c^{-1}) \right) \right] \\
&= \left[\left((1 + c) \times (1 + c^{-1}) \right) + 1, (c + 1) + (c^{-1} + 1) \right] \\
&= \left[(1 + 1) + \left(1 + (c + c^{-1}) \right), 1 + \left(1 + (c + c^{-1}) \right) \right] \quad \bigstar \\
&= [1 + 1, 1] = \mathbf{1}
\end{aligned}
$$

where I leave it to the reader to perform the calculations required to get to $\bigstar$ from the previous line.

If $a' > a$, then the previous two paragraphs tell us that $\tilde{\mathbf{a}} = [a', a]$ has a multiplicative inverse $\tilde{\mathbf{a}}^{-1} = [b', b]$, say. Writing $\mathbf{a}^{-1} = [b, b']$, we have, using Exercise 3.2.6,

$$
\mathbf{a} \otimes \mathbf{a}^{-1} = [a, a'] \otimes [b, b'] = [a', a] \otimes [b', b] = \tilde{\mathbf{a}} \otimes \tilde{\mathbf{a}}^{-1} = \mathbf{1}.
$$

(xiii) We have $a + b' > a' + b$ and $c > c'$, so (using the multiplication law for inequalities) $(a + b') \times c > (a' + b) \times c$ and $(a' + b) \times c > (a' + b) \times c'$. Thus, by the transitivity of order for the strictly positive rationals, $(a + b') \times c > (a' + b) \times c'$. ∎

Exercise 3.2.8 *Do the calculations required to obtain $\bigstar$ in the proof above.*

Exercise 3.2.9 *Prove the remaining parts of the theorem.*

Exercise 3.2.10 *By giving an explicit example, show that $(xiii)$ is false if we omit the condition $\mathbf{c} \ominus \mathbf{0}$.*

Exercise 3.2.11 *Consider any system satisfying the conclusions of Theorem 3.2.7. Prove the following results and briefly explain what they mean.*

(i) If $\tilde{\mathbf{0}} \oplus \mathbf{a} = \mathbf{a}$ for all $\mathbf{a}$, then $\tilde{\mathbf{0}} = \mathbf{0}$.
(ii) If $\mathbf{a} \oplus \mathbf{a}^\bullet = \mathbf{0}$, then $\mathbf{a}^\bullet = -\mathbf{a}$.
(iii) If $\tilde{\mathbf{1}} \otimes \mathbf{a} = \mathbf{a}$ for all $\mathbf{a}$, then $\tilde{\mathbf{1}} = \mathbf{1}$.
(iv) If $\mathbf{a} \neq \mathbf{0}$ and $\mathbf{a} \otimes \mathbf{a}^* = \mathbf{1}$, then $\mathbf{a}^* = \mathbf{a}^{-1}$.

[Look at Lemma 2.3.4 if you need a hint.]

We sometimes refer to $\mathbf{0}$ as an *additive zero* and call $-\mathbf{a}$ an *additive inverse*.

Exercise 3.2.12 *Consider any system satisfying the conclusions of Theorem 3.2.7. By considering inverses, show that it obeys the following cancellation laws.*

(i) *If* $\mathbf{a} \oplus \mathbf{c} = \mathbf{b} \oplus \mathbf{c}$, *then* $\mathbf{a} = \mathbf{b}$.
(ii) *If* $\mathbf{a} \otimes \mathbf{c} = \mathbf{b} \otimes \mathbf{c}$, *and* $\mathbf{c} \neq \mathbf{0}$, *then* $\mathbf{a} = \mathbf{b}$.
(iii) *If* $\mathbf{a}$ *and* $\mathbf{b}$ *are given, there exists a* $\mathbf{c}$ *such that* $\mathbf{a} \oplus \mathbf{c} = \mathbf{b}$.
(iv) *If* $\mathbf{a}$ *and* $\mathbf{b}$ *are given, and* $\mathbf{a} \neq \mathbf{0}$, *there exists a* $\mathbf{c}$ *such that* $\mathbf{a} \otimes \mathbf{c} = \mathbf{b}$.

Exercise 3.2.13 *Consider any system satisfying the conclusions of Theorem 3.2.7 for addition and order. If* $\mathbf{a} \ominus \mathbf{0}$, *show that* $\mathbf{0} \ominus -\mathbf{a}$ *and, if* $\mathbf{0} \ominus \mathbf{b}$, *then* $-\mathbf{b} \ominus \mathbf{0}$.

Exercise 3.2.14 *Consider any system satisfying the conclusions of Theorem 3.2.7.*

(i) *Show that* $-(-\mathbf{a}) = \mathbf{a}$.
(ii) *By applying the distributive law to* $\mathbf{a} \otimes (\mathbf{1} \oplus \mathbf{0})$, *show that* $\mathbf{a} \otimes \mathbf{0} = \mathbf{0}$ *for all* $\mathbf{a}$.
(iii) *By applying the distributive law to* $\mathbf{a} \otimes \big(\mathbf{b} \oplus (-\mathbf{b})\big)$, *show that* $(-\mathbf{a}) \otimes \mathbf{b} = -(\mathbf{a} \otimes \mathbf{b})$.
(iv) *By using (i) and (iii), show that* $(-\mathbf{a}) \otimes (-\mathbf{b}) = \mathbf{a} \otimes \mathbf{b}$.

[*This exercise makes precise our previous assertion that* only *the multiplication rules of Brahmagupta allow our extended system to work as we wish. Notice that you need the uniqueness results of Exercise 3.2.11 to obtain the results of this exercise.*]

Exercise 3.2.15 *We continue with the ideas of Exercise 3.2.14.*

(i) *Show that* $-(\mathbf{a} \oplus \mathbf{b}) = (-\mathbf{a}) \oplus (-\mathbf{b})$.
(ii) *If* $\mathbf{a}$, $\mathbf{b} \neq \mathbf{0}$, *show that* $\mathbf{a} \otimes \mathbf{b} \neq \mathbf{0}$ *and* $(\mathbf{a} \otimes \mathbf{b})^{-1} = \mathbf{a}^{-1} \otimes \mathbf{b}^{-1}$.
(iii) *(This complements the footnote on page 44.) Show that*

$$(\mathbf{a} \oplus (-\mathbf{a}')) \otimes (\mathbf{b} \oplus (-\mathbf{b}')) = ((\mathbf{a} \otimes \mathbf{b}) \oplus (\mathbf{a}' \otimes \mathbf{b}')) \oplus (-((\mathbf{a}' \otimes \mathbf{b}) \oplus (\mathbf{a} \otimes \mathbf{b}'))).$$

Exercise 3.2.16 *Consider any system satisfying the conclusions of Theorem 3.2.7.*

(i) *By using Exercise 3.2.14 (iv), or otherwise, show that* $(-\mathbf{1}) \otimes (-\mathbf{1}) = \mathbf{1}$.
(ii) *If* $\mathbf{a} \ominus \mathbf{0}$, *show that* $\mathbf{a} \otimes \mathbf{a} \ominus \mathbf{0}$.
(iii) *Show that, if* $\mathbf{a} \neq \mathbf{0}$, *then* $\mathbf{a} \otimes \mathbf{a} \ominus \mathbf{0}$.

(iv) By giving explicit examples, show that part (ii) of Exercise 3.2.12 is false if
we omit the condition **c** $\neq$ **0** *and part (iv) is false if we omit the condition*
a $\neq$ **0**.
(v) Show that **1** $\ominus$ **0** *and deduce that* **0** $\ominus$ **−1**.
(vi) Show that the equation

$$\mathbf{a} \otimes \mathbf{a} = -\mathbf{1}$$

has no solution.
[As the reader probably expects, part (vi) will be important when we come to
look at complex numbers.]

Our experience with the extension of the natural numbers to the strictly
positive rationals leads us to expect the following companion to Lemma 2.3.7

Exercise 3.2.17 *By setting* $f(a) = [a + 1, 1]$, *show that there is an injective*
function $f : \mathbb{Q}^+ \to B/\sim$ *such that*

$$f(a + b) = f(a) \oplus f(b), \ f(a \times b) = f(a) \otimes f(b)$$
$$\text{and } a > b \text{ implies } f(a) \ominus f(b).$$

Just as we eventually set $\mathbb{Q}^+ = A/\sim$ so now we set $\mathbb{Q} = B/\sim$ and replace
our fancy $\oplus$ with $+$ and so on.
We constructed $\mathbb{Q}^+$ from $\mathbb{N}^+$ and $\mathbb{Q}$ from $\mathbb{Q}^+$.

Lemma 3.2.18 *There is an injective function* $F : \mathbb{N}^+ \to \mathbb{Q}$ *such that*

$$F(n + m) = F(n) + F(m), \ F(n \times m) = F(n) \times F(m)$$
$$\text{and } n > m \text{ implies } F(n) > F(m).$$

Proof. By Lemma 2.3.7, there is an injective function $f_1 : \mathbb{N}^+ \to \mathbb{Q}^+$ such
that

$$f_1(n + m) = f_1(n) + f_1(m), \ f_1(n \times m) = f_1(n) \times f_1(m)$$
$$\text{and } n > m \text{ implies } f_1(n) > f_1(m).$$

By Exercise 3.2.17 there is an injective function $f_2 : \mathbb{Q}^+ \to \mathbb{Q}$ such that

$$f_2(a + b) = f_2(a) + f_2(b), \ f_2(a \times b) = f_2(a) \times f_2(b)$$
$$\text{and } a > b \text{ implies } f_2(a) > f_2(b).$$

If we set $F(n) = f_2(f_1(n))$ we obtain the desired result. ∎

We have shown that the natural numbers may be considered as a subset of $\mathbb{Q}$ by identifying n with $f(n)$, but we have not yet defined the integers in general.

Definition 3.2.19 *We write $\mathbb{Z}$ for the subset of $\mathbb{Q}$ consisting of the elements of the form $n - m$ such that n and m lie in $\mathbb{N}^+$.*

We call the members of $\mathbb{Z}$, *integers*.

Exercise 3.2.20 *Prove the following results about $\mathbb{Z}$.*

(i) If $a \in \mathbb{N}^+$, then $a \in \mathbb{Z}$.
(ii) If $a, b \in \mathbb{Z}$, then $a + b$, $a \times b$, $-a \in \mathbb{Z}$.

When we work in $\mathbb{Z}$ we shall consider the two terms 'natural numbers' and 'strictly positive integers' as synonymous. When we talk about the 'non-negative integers'[9] we shall mean integers n with $n \geq 0$.

We shall often use the customary notations $a - b = a + (-b)$ and

$$a \div b = a/b = a \times b^{-1},$$

but we shall not treat the four branches of arithmetic[10] as being on the same footing for the reasons given in the next very short exercise.

Exercise 3.2.21 *Without worrying about mathematical niceties, just write down for each part, n, m and p in $\mathbb{Q}$ satisfying the following conditions.*

(i) $n - m \neq m - n$.
(ii) $n \div m \neq m \div n$.
(iii) $n - (m - p) \neq (n - m) - p$.
(iv) $n \div (m \div p) \neq (n \div m) \div p$.

Thus subtraction and division are neither commutative nor associative.

---

[9] Many mathematicians refer to the set $\mathbb{N}$ of non-negative integers as the natural numbers, so that 0 is a natural number. Some mathematicians hold strong views on the matter. Others are content to quote Byrom.

> Some say, compar'd to Bononcini,
> That Mynheer Handel 's but a ninny;
> Others aver that he to Handel
> Is scarcely fit to hold a candle.
> Strange all this difference should be
> 'Twixt Tweedledum and Tweedledee.

[10] 'The different branches of Arithmetic – Ambition, Distraction, Uglification, and Derision.' Lewis Carroll, *Alice in Wonderland.*

3.3 What Does Nature Say?

So far our examples of 'real life' use of negative numbers (debts, temperature scales, ...) have involved addition. The only people to multiply negative numbers during the first thousand years of their existence were the puzzlers (or pure mathematicians, if you prefer), although, towards the end of the period, the puzzles became more complicated. (In 1545 Cardano announced the solution of the general cubic by del Fero and of the general quartic equation by Ferrari[11].)

It was thus an extraordinary event when Galileo showed that the flight of a cannon ball was governed by the same mathematics as Bhāskarāchārya's troops of monkeys. Henceforward, the multiplication of negative numbers would be a matter for serious-minded military engineers.

Newton presented his mechanics to the world in geometric form, but his successors used the flexibility of negative numbers to develop mechanics further. Later physical theories such as electromagnetism, relativity and quantum mechanics have negative numbers and their multiplication rules built into their deepest structure. Every time we see an experimental verification of General Relativity we see a test of the applicability of Brahmagupta's multiplication rules.

Should we conclude that the multiplication laws are built into nature[12]? Suppose that we have two cannon balls falling vertically, the first having velocity $v_1(t)$ and the second velocity $v_2(t)$ at time t. If $v_2(s) = v_1(0)$ (that is to say, the second cannon ball has the same velocity as the first had at time 0, but at time s) we expect $v_2(t + s) = v_1(t)$, that is to say, the second cannon ball behaves like the first but 'shifted in time' by an amount s. More generally we expect our physical theories to be 'translation invariant' in time.

Galileo's law for falling bodies may be expressed as

$$v(t) = C + (g \times t),$$

where $v(t)$ is the velocity at time t and C is a constant. Thus

$$v_1(t) = A + (g \times t) \text{ and } v_2(t) = B + (g \times t)$$

for appropriate constants A and B. It follows that

$$B + \big(g \times (t + s)\big) = A + (g \times t).$$

---

[11] A cubic equation has the form $x^3 + ax^2 + bx + c = 0$ and a quartic equation has the form $x^4 + ax^3 + bx^2 + cx + d = 0$.

[12] In the next three paragraphs the author ventures far out of his depth. The reader should feel free to ignore everything in them.

Taking $t = 0$, using the natural definition for multiplication by zero and various laws for the behaviour of addition, we get

$$A = B + (g \times s).$$

Thus

$$B + \big((g \times s) + (g \times t)\big) = \big(B + (g \times s)\big) + (g \times t) = A + (g \times t) = B + \big(g \times (t+s)\big),$$

so, cancelling,

$$(g \times s) + (g \times t) = g \times (s + t).$$

Thus the distributive law is, in some sense, implicit in Galileo's law. The distributive law was our main tool in Exercise 3.2.14 for showing that Brahmagupta's rules are the only ones which give the negative numbers the properties we desire.

It appears that 'translation invariant theories' demand 'translation invariant arithmetic' and 'translation invariant arithmetic' requires Bramahgupta's rules. A more modest version of the statement (and, I think, a more correct one) would be that writing a theory in 'translation invariant arithmetic' allows an easy check that the theory is indeed 'translation invariant' in the appropriate sense. However, such speculations are idle unless considered by an Einstein or an Emmy Noether.

It is more important to remember that nature is not bound by our arithmetic. Although we have used temperature scales as an example of an extended number system, deeper study reveals the existence of an 'absolute zero' so that 'natural' temperature scales such as those in degrees Kelvin correspond to the positive numbers only. In modern physics, the positron is, in effect, a 'negative' electron. However, the addition of a positron to an electron does not produce nothingness, but a burst of energy.

3.4 When Are Two Things the Same?

The ideas of this section occur in every university mathematics course, so the reader may choose merely to skim across the discussion.

We begin with Anne Duncan's game[13].

Example 3.4.1 *This game requires nine cards on each of which is written a different word from the sentence*

---

[13] A variation on Leo Moser's game taken from the treasure chest *Winning Ways for Your Mathematical Plays* [3].

SPIT NOT SO, FAT FOP, AS IF IN PAN!

The two players pick cards alternately and the player (if any) who first gets a set of three words with the same letter (for example: SPIT, SO, AS) wins.

The reader should spend a few minutes thinking about how to play this game.

Since this section is about things being the same, she may also wish to reflect on the following paradox of Eubulides[14] called the *paradox of the masked man.*

Example 3.4.2 *Consider the following conversation:*
  'Do you know this masked man?'
  'No.'
  'Do you know your father?'
  'Of course.'
  'But the masked man is your father.'

We constructed (a version of) $\mathbb{Q}$ from $\mathbb{N}^+$ by first constructing (a version of) $\mathbb{Q}^+$ from $\mathbb{N}^+$ and then constructing (a version of) $\mathbb{Q}$ from (our version of) $\mathbb{Q}^+$. But we could equally well have first constructed (a version of) $\mathbb{Z}$, that is to say, the integers and then constructed (a version of) $\mathbb{Q}$ from our new version of $\mathbb{Z}$.

Exercise 3.4.3 *(Just make sure you have the general idea of what is going on.) By imitating our construction of (a version of) $\mathbb{Q}$ from (a version of) $\mathbb{Q}^+$ show how to construct (a version of) $\mathbb{Z}$ from $\mathbb{N}^+$ by using ordered pairs. Call your version $\mathcal{A}$. Write down the properties that your system has. (It is an ordered integral domain, so you may check your list against Definitions 10.4.1 and 10.4.7.) Do not do the calculations to verify the properties unless you are very interested. (The most interesting check is that of the multiplicative cancellation law for $\mathbb{Z}$: if $n \times m = 0$, then $n = 0$ or $m = 0$.)*

Now, by imitating our construction of (a version of) $\mathbb{Q}^+$ from (a version of) $\mathbb{N}^+$, show how to construct (a version of) $\mathbb{Q}$ from $\mathcal{A}$ by using ordered pairs. Do not do the calculations to verify the properties unless you are very interested.

Suppose now that mathematician *A* constructs her 'rational number system' using our first method and mathematician *B* constructs her 'rational number

---

[14] Eubulides was the author of four splendid paradoxes. These include, in addition to the paradox of the masked man, the *paradox of the liar* which finds echoes in Example 6.5.5. One version of this paradox is the sentence: 'This statement is false.'

system' using the method just outlined. Are they the same? If two cooks bake chocolate cakes using different recipes, are they the same?

When faced with a deep philosophical problem, the mathematician's response, like that of an invading army faced with a fortress, is to bypass it rather than besiege it. Rather than seeking to resolve the problem, mathematicians seek to evade it.[15] To see how we might do this in the present case, let us return to Anne Duncan's game.

The nature of the game becomes clearer if we arrange the words in an array as follows.

NOT	IN	PAN
SO	SPIT	AS
FOP	IF	FAT

If the first player replaces any card she takes with a cross X and the second player replaces any card she takes with a nought 0, it soon becomes clear that the game they are playing is 'Noughts and Crosses' or (if they are on the other side of the Atlantic) 'Tic-Tac-Toe'.[16]

Can we say that Anne Duncan's game is Noughts and Crosses? Clearly not. Noughts and Crosses is a game played with pencil and paper, while Anne Duncan's game is a game played with nine cards. What we can say is that there is an exact translation from the language of one to the language of the other when deciding the legality of moves and the result of the game. Informally, the two games are identical if we are interested only in the legality of a given game between two players and in the result. Mathematicians do not ask if two systems are identical but if two systems are *identical for particular purposes*. If the two systems are identical for a particular purpose, they say that the systems are *isomorphic for that purpose*.

Exercise 3.4.4 *In the game* Magic Fifteen,[17] *two players alternately select integers r with $1 \leq r \leq 9$ with no integer being used twice. You win by getting three numbers whose sum is* 15. *Comment.*

[*It may be helpful to write out the winning combinations.*]

We defined the notion of an injective function in Definition 2.3.6 which the reader should reread. In order to make the notion of an isomorphism more precise, we need some further definitions along the same lines.

---

[15] We shall see further examples of this in Section 6.5.

[16] Where the player who succeeds in placing three of her marks in a horizontal, vertical or diagonal row wins the game.

[17] Also taken from *Winning Ways for Your Mathematical Plays* [3], where it is attributed to E. Pericoloso Sporgersi.

Definition 3.4.5 *We say that a function* $f : X \to Y$ *(that is to say, a function taking members of X to members of Y) is* surjective *if, whenever* $y \in Y$, *there exists an* $x \in X$ *such that* $f(x) = y$.

Definition 3.4.6 *We say that a function* $f : X \to Y$ *is* bijective *if f is both injective and surjective.*

It will be useful to give an *explicit* definition for the equality of functions (which will, I hope, agree with what the reader expects).

Definition 3.4.7 *If* $f : X \to Y$, $g : X \to Y$ *are functions, we say that* $f = g$ *if* $f(x) = g(x)$ *for all* $x \in X$.

Exercise 3.4.8 *Let* $X = \{1, 2\}$ *(that is to say, let X be the set with members 1 and 2) and let* $Y = \{1, 2, 3\}$. *For each of the following functions, state, with brief reasons, whether it is injective, whether it is surjective and whether it is bijective.*

(i) $f_1 : X \to X$ *with* $f_1(1) = f_1(2) = 1$.
(ii) $f_2 : X \to X$ *with* $f_2(1) = 2$, $f_2(2) = 1$.
(iii) $f_3 : X \to Y$ *with* $f_3(1) = 1$, $f_3(2) = 2$.
(iv) $f_4 : Y \to X$ *with* $f_4(1) = 1$, $f_4(2) = 2$, $f_4(3) = 1$.

Exercise 3.4.9 *For each of the following functions, state, with brief reasons, whether it is injective, whether it is surjective and whether it is bijective.*

(i) $f_1 : \mathbb{Z} \to \mathbb{Z}$ *with* $f_1(r) = 2 \times r$.
(ii) $f_2 : \mathbb{Q} \to \mathbb{Q}$ *with* $f_2(r) = 2 \times r$.
(iii) $f_3 : \mathbb{Z} \to \mathbb{Z}$ *with* $f_3(2 \times r) = r$, $f_3\big((2 \times r) + 1\big) = r$.
(iv) $f_4 : \mathbb{Z} \to \mathbb{Z}$ *with* $f_4(r) = r \times r$.

We make a couple of simple, but important, observations.

Theorem 3.4.10

(i) *If* $f : X \to Y$ *is bijective, then there exists a unique function* $g : Y \to X$ *such that* $f\big(g(y)\big) = y$ *for all* $y \in Y$.
(ii) *The function* $g : Y \to X$ *defined in (i) is bijective and* $g\big(f(x)\big) = x$ *for all* x.

We write $f^{-1} = g$ and call g the *inverse* of f.

Proof of Theorem 3.4.10. (i) We first prove existence. If $y \in Y$, then, since f is injective and surjective, there exists a unique $z \in X$ such that $f(z) = y$. We write $g(y) = z$.

We now prove uniqueness. If $g_i : Y \to X$ satisfies $f\big(g_i(y)\big) = y$ for each $y \in Y$ [$i = 1, 2$], then

$$f\big(g_1(y)\big) = f\big(g_2(y)\big)$$

for all $y \in Y$. Since f is injective, we have

$$g_1(y) = g_2(y)$$

for all $y \in Y$ and so $g_1 = g_2$.

(ii) If $x \in X$, then $f(x) \in Y$, so

$$f\big(g(f(x))\big) = f(x).$$

Since f is injective, $g(f(x)) = x$ for all $x \in X$. Automatically, g is surjective.

Next we show g injective. If $g(y_1) = g(y_2)$, then

$$y_1 = f\big(g(y_1)\big) = f\big(g(y_2)\big) = y_2;$$

so we are done. ∎

Exercise 3.4.11 *Consider two functions $f : X \to Y$ and $g : Y \to X$.*

(i) *If $f\big(g(y)\big) = y$ for all $y \in Y$, show that f is surjective and g is injective.*

(ii) *If $f\big(g(y)\big) = y$ for all $y \in Y$ and $g\big(f(x)\big) = x$ for all $x \in X$, show that f is bijective and $g = f^{-1}$.*

(iii) *Find X, Y, f and g such that $f\big(g(y)\big) = y$ for all $y \in Y$, but g is not surjective.*

(iv) *Find X, Y, f and g such that $f\big(g(y)\big) = y$ for all $y \in Y$, but f is not injective.*

Once we have the idea of a bijection, we can see what should be meant by isomorphism.

Definition 3.4.12 *(Note that this is not really a definition, but a blueprint for a series of definitions.) If A and B are sets with associated operations and relations we say that the associated systems are* isomorphic *if there exists a bijection $f : A \to B$ which preserves the operations and relations. We call f an* isomorphism.

Let us see how this works in the case of a single operation.

Definition 3.4.13 *Let A and B be sets with associated operations $+$ and $\oplus$. We say that $(A, +)$ is* isomorphic *to $(B, \oplus)$ if there exists a bijection $f : A \to B$ with $f(x + y) = f(x) \oplus f(y)$ for all x, $y \in A$. We call f an* isomorphism.

Here is a typical result on isomorphisms.

Lemma 3.4.14 *Let A, B and C be sets with associated operations $+$, $\oplus$ and $\boxplus$. Let us write $(A, +) \sim (B, \oplus)$ to mean $(A, +)$ is isomorphic to $(B, \oplus)$ and so on. The following results hold.*

(i) If $(A, +) \sim (B, \oplus)$, then $(B, \oplus) \sim (A, +)$.

(ii) $(A, +) \sim (A, +)$.

(iii) If $(A, +) \sim (B, \oplus)$ and $(B, \oplus) \sim (C, \boxplus)$, then $(A, +) \sim (C, \boxplus)$.

Proof. We prove (i) and leave the rest to the reader. If $(A, +) \sim (B, \oplus)$ then, by definition, there exists a bijection $f : A \to B$ with $f(x+y) = f(x) \oplus f(y)$ for all x, $y \in A$. We observe that f^{-1} is then a bijection and

$$f(f^{-1}(u) + f^{-1}(v)) = f\big(f^{-1}(u)\big) \oplus f\big(f^{-1}(v)\big) = u \oplus v = f\big(f^{-1}(u \oplus v)\big);$$

so, since f is injective,

$$f^{-1}(u) + f^{-1}(v) = f^{-1}(u \oplus v)$$

for all u, $v \in B$. Thus f^{-1} is an isomorphism. ∎

Exercise 3.4.15 *Prove the remaining parts of Lemma 3.4.14.*

Example 3.4.16 *Let $\mathbb{N}^+$ be the set of natural numbers with the usual addition $+$ and multiplication $\times$ and let E be the set of strictly positive even integers with the same operations now denoted by $\otimes$ and $\oplus$. Then the following results hold.*

(i) $(\mathbb{N}^+, +)$ is isomorphic to $(E, \oplus)$.

(ii) $(\mathbb{N}^+, \times)$ is not isomorphic to $(E, \otimes)$.

Proof. (i) Let us write the elements of E as $2n = 2 \times n$. Consider $f : \mathbb{N}^+ \to E$ given by $f(n) = 2 \times n$. The function f is a bijection with

$$f(n + m) = 2(n + m) = 2n + 2m = f(n) \oplus f(m).$$

(ii) Suppose that $f : \mathbb{N}^+ \to E$ is a bijection which preserves multiplication. Let $f(1) = 2N$. Then

$$2N = f(1) = f(1 \times 1) = f(1) \otimes f(1) = (2N) \times (2N);$$

so $1 = 2N$, which is not possible. ∎

Not surprisingly, it turns out that the various constructions of models for the rationals all give the same answer: 'up to the appropriate isomorphism'. In Exercise 7.2.8 we shall see the reason (or, at least, a very good reason) why this is so.

Remark Isomorphism is a relation between systems. If we work within a system like the natural numbers, then the same thing may have different names. Thus 5, $2 + 3$, $3 + 2$ and 'the natural number whose square is 25' are just different names for the same thing. (Compare the paradox of Eubulides in Exercise 3.4.2.) When we write $a = b$ we mean that a and b are both names for the same thing.

PART II

The Natural Numbers

4

The Golden Key

4.1 The Least Member

We have written down rules for addition, multiplication and order for the natural numbers, but the result is strangely lifeless. Either the strictly positive integers are as dull as we have made them seem or there is some property that they have which we have failed to observe.

Fortunately there is a further property, which seems to have first been recognised explicitly by Fermat. We shall call it the *least member principle*. It is the golden key to the understanding of the natural numbers.

Least member principle *A non-empty[1] collection E of natural numbers always has a least member.*

More formally, if E is a non-empty collection of natural numbers, then we can find $e_0 \in E$ such that $e \geq e_0$ for all $e \in E$.

In more advanced work the least member principle becomes the statement that '$\mathbb{N}^+$ is *well ordered* by $>$'.

We need to make a simple remark.

Lemma 4.1.1 *If E is a non-empty collection of natural numbers, then it has a unique least member.*

Proof. In view of the least member principle, we need only show uniqueness. If e_0, $e_1 \in E$ are such that $e \geq e_0$ and $e \geq e_1$ for all $e \in E$, then, in particular, $e_1 \geq e_0$ and $e_0 \geq e_1$; so, by trichotomy, $e_0 = e_1$. ∎

Because of this uniqueness result, we may talk about *the* least member.

---

[1] We say that a collection is *empty* if it has no members. Thus the collection of square circles is empty. This convention is useful because it allows us to talk about collections of objects with property P even if we do not know if any such objects actually exist. As might be expected, we say that a collection is *non-empty* if it is not empty.

Exercise 4.1.2

(i) If we consider $\mathbb{Z}$ with the usual order $>$, show that $\mathbb{Z}$ has no least member.
(ii) If we consider $\mathbb{Q}$ with the usual order, show that the collection E of $q \in \mathbb{Q}$ with $q > 1$ has no least member.

Even if the least member principle is unfamiliar to the reader, she will probably know the principle of induction[2] (usually attributed to Pascal).

Principle of induction *Let $P(n)$ be the statement that a natural number n has a property P. If $P(1)$ is true and, whenever $P(n)$ is true, so is $P(n + 1)$, then $P(n)$ is true for all n.*

The reader will also, consciously or not, be familiar with another principle (which would certainly be accepted by any farmer when counting sheep).

Base number principle[3] *If n is a natural number, then $n \geq 1$.*

Exercise 4.1.3 *Using the subtraction principle, show that the base number principle implies that, if r, $n \in \mathbb{N}^+$ with $r > n$ and $n + 1 \geq r$, then $r = n + 1$.*

In this section we shall show that the two principles just given are *together* equivalent to the least member principle.

We start by deducing the base number principle from the least member principle.

Theorem 4.1.4 *The rules we have given for the natural numbers, together with the least member principle, imply that 1 is the least natural number.*

Proof. The least member principle tells us that $\mathbb{N}^+$ has a least member; call it e_0. By definition $1 \geq e_0$. Suppose, if possible, that $1 > e_0$. The multiplication law for inequalities (see Lemma 1.3.8) tells us that, if $a > b$, then $a \times c > b \times c$. Choosing $a = 1$ and $b = e_0$, we obtain

$$e_0 = 1 \times e_0 > e_0 \times e_0.$$

Since e_0 is the least element of $\mathbb{N}^+$, this gives a contradiction.

It follows that $e_0 \geq 1$ and so, since $1 \geq e_0$, trichotomy tells us that $1 = e_0$. ∎

Theorem 4.1.5 *The rules we have given for the natural numbers, together with the least member principle, imply the principle of induction.*

---

[2] Often called the principle of *mathematical induction* to distinguish it from other uses of the word induction.

[3] We use this name because $P(1)$ is often called the base case for an induction.

Proof. Let $P(n)$ be the statement that a natural number n has a property P. Suppose that $P(1)$ is true and, whenever $P(n)$ is true, so is $P(n+1)$. We wish to show that $P(n)$ is true for all n.

Let E be the collection of natural numbers such that $P(n)$ is false. If E is empty, then $P(n)$ is true for all n. Suppose, if possible, that E is non-empty. Then E has a least member e_0. Since $P(1)$ is true, $e_0 \neq 1$ and, since 1 is the least natural number (by Theorem 4.1.4), $e_0 > 1$. The order rule (see page 18) tells us that, if $a > b$, then we can find a natural number c such that $b + c = a$. Taking $a = e_0$ and $b = 1$, we see that there is a c such that $e_0 = 1 + c = c + 1$. Now e_0 is the least natural number e such that $P(e)$ is false, so $P(c)$ is true and, by hypothesis, $P(e_0) = P(c+1)$ is true.

Since the assumption that E is non-empty leads to a contradiction, we must have E empty and we are done. ∎

We need a preliminary lemma along the lines of Exercise 4.1.3 before we prove the converse.

Lemma 4.1.6 *The rules we have given for the natural numbers, together with the base number principle, imply that, if r and n are natural numbers with $n + 1 > r$, then $n \geq r$.*

Proof. Suppose, if possible, that it is not true that $n \geq r$ and so, by trichotomy, $r > n$. By the order rule (see page 18), we can find a natural number c such that $r = n + c$. Since $n + 1 > r$, it follows that $n + 1 > n + c$ and, by the cancellation law for addition (Lemma 1.3.9 (i)), $1 > c$. This contradicts the base number principle, so our initial assumption must be false. ∎

Theorem 4.1.7 *The rules we have given for the natural numbers, together with the principle of induction and the base number principle, imply the least member principle.*

Proof. Assume the principle of induction and the base number principle. Let $P(n)$ be the statement that any collection E of natural numbers containing a natural number r with $r \leq n$ has a least member.

If E is a collection of natural numbers containing 1, then the base number principle tells us that $e \geq 1$ for all $e \in E$, so 1 is the least element of E. Thus $P(1)$ is true.

Now suppose that $P(n)$ is true. If E is a collection of natural numbers containing a natural number $r \leq n + 1$, then either $n + 1$ is the only member of E and so $n + 1$ is certainly the least member of E, or E contains an s with $n + 1 > s$. In the second case, Lemma 4.1.6 tells us that $n \geq s$ and $P(n)$ now tells us that E has a least member.

We have shown that $P(n)$ implies $P(n+1)$ and that $P(1)$ is true. The principle of induction tells us that $P(n)$ is true for all n. Since any non-empty collection of natural numbers E contains *some* number m, we know that, since $P(m)$ is true, E has a least member. ∎

The following remark is clearly going to be useful.

Lemma 4.1.8 *The rules we have given for the natural numbers together with the least member principle imply that, if E is a non-empty collection of natural numbers and we can find an n with $n \geq e$ for all $e \in E$, then there exists an $e_1 \in E$ such that $e_1 \geq e$ for all $e \in E$.*

Informally, any non-empty collection of natural numbers which is bounded above has a greatest member.

Exercise 4.1.9

(i) *Show that the greatest member, just referred to, is unique.*
(ii) *Give an example of a non-empty collection of natural numbers with no greatest member.*

Proof of Lemma 4.1.8. Suppose that E is a non-empty collection of natural numbers and we can find an n with $n \geq e$ for all $e \in E$. If $n \in E$, then n is the greatest member of E. If not, then, if $e \in E$, we know that $n > e$ and so, by the order rule, we can find a natural number $c(e)$ such that $e + c(e) = n$. The collection of $c(e)$ has a least member $c(e_1)$ with $e_1 \in E$. Thus, whenever $e \in E$,

$$e_1 + n = e_1 + \left(e + c(e)\right) = (e_1 + e) + c(e)$$
$$\geq (e_1 + e) + c(e_1) = e + \left(e_1 + c(e_1)\right) = e + n,$$

and so, by cancellation (see Lemma 1.3.9 (i)), $e_1 \geq e$. ∎

Exercise 4.1.10 *Use the least member principle and the base number principle to prove the following version of the principle of induction. Let $Q(n)$ be the statement that a natural number n has a property Q. If $Q(1)$ is true and, whenever $Q(r)$ is true for all $1 \leq r \leq n$, so is $Q(n+1)$, then $Q(n)$ is true for all $n \in \mathbb{N}^+$.*

Exercise 4.1.11 *We work in $\mathbb{N}^+$ and take some $m \in \mathbb{N}^+$. Suppose that $P(n)$ implies $P(n+1)$ for all $n \geq m$ and that $P(m)$ is true. Show that $P(n)$ is true for all $n \geq m$.*

We conclude this section with a result which is extremely useful when discussing repetitive calculations, since it gives a method for showing that such calculations will stop. We will illustrate its use in the next section when we discuss Euclid's algorithm.

Lemma 4.1.12 *Any decreasing sequence[4] of natural numbers is eventually constant. In other words, if* $n_j \in \mathbb{N}^+$ $[j \geq 1]$ *and*

$$n_1 \geq n_2 \geq n_3 \geq \ldots,$$

then there exists an N such that $n_j = n_N$ *for all* $j \geq N$.

Proof. The collection of n_j must have a least member, n_N. If $n_j = n_N$ then, since $n_j \geq n_{j+1}$ transitivity of order tells us that $n_N \geq n_{j+1}$ and so, since $n_{j+1} \geq n_N$, trichotomy tells us that $n_{j+1} = n_N$. The form of induction given in Exercise 4.1.11 now tells us that $n_j = n_N$ for all $j \geq N$. ∎

Exercise 4.1.13

(i) *Show that there does not exist a strictly decreasing sequence of natural numbers. In other words, we cannot find* $n_j \in \mathbb{N}^+$ $[j \geq 1]$ *such that*

$$n_1 > n_2 > n_3 > \ldots.$$

(ii) *Give an example of a strictly decreasing sequence of integers.*

4.2 Inductive Definition

The use of induction is not restricted to proof. We can also use it to define new objects.

Informal notion of inductive definition[5] *If we define* $f(1)$ *and have a procedure which, given* $f(n)$, *defines* $f(n+1)$, *then* $f(n)$ *is defined for all n.*

The notion of 'a procedure' is rather vague, so we use the following theorem which, although it may not capture the full flavour of our informal statement, is sufficient for the purposes of this book.

Theorem 4.2.1 *Let X be a set with* $x_1 \in X$. *Suppose that whenever* $n \in \mathbb{N}^+$ *we have a function* $g_n : X \to X$. *Then there exists a unique* $f : \mathbb{N}^+ \to X$ *such that* $f(1) = x_1$ *and* $f(n+1) = g_n(f(n))$.

---

[4] Unless otherwise stated, sequence will always mean what is sometimes called 'an infinite sequence'.

[5] Logicians and computer scientists prefer to use the term 'recursive definition'.

The following lemma provides a stepping stone towards the proof of Theorem 4.2.1.

Lemma 4.2.2 *Suppose that $m \in \mathbb{N}^+$ and hypotheses of Theorem 4.2.1 apply. Then, if S_m is the collection of $r \in \mathbb{N}^+$ with $r \leq m$, there exists a unique function $f_m : S_m \to X$ such that $f(1) = x_1$ and $f(n + 1) = g_n(f(n))$ whenever $n \in S_m$ and $n + 1 \leq m$.*

Proof. (This proof and the next one make repeated use of Exercise 4.1.3.) We first prove uniqueness. Let $P(m)$ be the statement that if $h, k : S_m \to X$ satisfy $h(1) = k(1) = x_1$ and $h(n + 1) = g_n(h(n))$, $k(n + 1) = g_n(k(n))$ whenever $n + 1 \leq m$, then $h(n) = k(n)$ for all $n \in S_m$. We observe that $P(1)$ is automatically true since there does not exist any $n \in S_1$ with $n + 1 \leq 1$.

Now suppose that $P(m)$ is true and $h, k : S_{m+1} \to X$ with $h(1) = k(1) = x_1$ and $h(n + 1) = g_n(h(n))$, $k(n + 1) = g_n(k(n))$ whenever $n + 1 \leq m + 1$. If we define $\tilde{h}, \tilde{k} : S_m \to X$ by $\tilde{h}(n) = h(n)$, $\tilde{k}(n) = k(n)$ for $n \in S_m$, then $\tilde{h}(1) = \tilde{k}(1) = x_1$ and $\tilde{h}(n + 1) = g_n(\tilde{h}(n))$, $\tilde{k}(n + 1) = g_n(\tilde{k}(n))$ whenever $n + 1 \leq m$. Since $P(m)$ is true we have $h(n) = \tilde{h}(n) = \tilde{k}(n) = k(n)$ for $n \in S_m$. It now follows that

$$h(m + 1) = g_m(h(m)) = g_m(k(m)) = k(m + 1),$$

so $h(n) = k(n)$ for all $n \in S_{m+1}$. We have shown that $P(m + 1)$ is true and the full result follows by induction.

We now prove existence. Let $Q(m)$ be the statement that there exists a function $f_m : S_m \to X$ such that $f_m(1) = x_1$ and $f_m(n+1) = g_n(f_m(n))$ whenever $n \in S_m$ and $n + 1 \leq m$. Setting $f_1(1) = x_1$, we see that $Q(1)$ is true. On the other hand, if $Q(m)$ is true we see, by setting

$$f_{m+1}(n) = \begin{cases} f_m(n) & \text{if } n \in S_m \\ g_m(f_m(m)) & \text{if } n = m + 1 \end{cases}$$

that $Q(m + 1)$ is true and the full result follows by induction. ∎

Once we have Lemma 4.2.2 the rest of the proof is straightforward.

Proof of Theorem 4.2.1. We first prove uniqueness. Suppose $h, k : \mathbb{N}^+ \to X$ are functions with $h(1) = k(1) = x_1$ and $h(n + 1) = g_n(h(n))$, $k(n + 1) = g_n(k(n))$ for all $n \in \mathbb{N}$.

Take any $m \in \mathbb{N}^+$. If we define $\tilde{h}, \tilde{k} : S_m \to X$ by $\tilde{h}(n) = h(n)$, $\tilde{k}(n) = k(n)$ for $n \in S_m$, then $\tilde{h}(1) = \tilde{k}(1) = x_1$ and

$$\tilde{h}(n + 1) = g_n(\tilde{h}(n)), \ \tilde{k}(n + 1) = g_n(\tilde{k}(n))$$

for all $n \in S_m$ with $n + 1 \leq m$. Lemma 4.2.2 now tells us that

$$h(n) = \tilde{h}(n) = \tilde{k}(n) = k(n)$$

for all $n \in S_m$ and so, in particular, $h(m) = k(m)$.

To prove existence, we observe, using the uniqueness part of Lemma 4.2.2 again, that $f_n(m) = f_p(m)$, whenever $m \leq n \leq p$. Thus writing $f(n) = f_n(n)$ we have $f(1) = f_1(1) = x_1$ and

$$f(n + 1) = f_{n+1}(n + 1) = g_n(f_n(n)) = g_n(f_n(n)) = g_n(f(n))$$

for all $n \in \mathbb{N}$ as required. ∎

Exercise 4.2.3 *Give the details of the proof of the statement that $f_n(m) = f_p(m)$, whenever $m \leq n \leq p$ in the first sentence of the last paragraph of the previous proof.*

We use an inductive definition to obtain a function $f_a(n)$, which the reader should recognise under its very light disguise.

Theorem 4.2.4 *If $a \in \mathbb{Q}$, then we define $f_a(n)$ inductively by setting $f_a(1) = a$ and $f_a(n + 1) = f_a(n) \times a$. Then, if $a, b \in \mathbb{Q}$ and $n, m \in \mathbb{N}^+$ the following results hold.*

(i) $f_a(m + n) = f_a(m) \times f_a(n)$ for all strictly positive integers m and n.
(ii) $f_a(n) \times f_b(n) = f_{a \times b}(n)$.
(iii) If $u(m) = f_a(m)$, then $f_{u(m)}(n) = f_a(m \times n)$.

Proof. (i) We show that the result is true for fixed m and any n. Fix m and let $P(n)$ be the statement that $f_a(m + n) = f_a(m) \times f_a(n)$. By definition,

$$f_a(m + 1) = f_a(m) \times a = f_a(m) \times f_a(1),$$

so $P(1)$ is true. If $P(n)$ is true, then

$$f_a(m + (n + 1)) = f_a((m + n) + 1) = f_a(m + n) \times a$$
$$= (f_a(m) \times f_a(n)) \times a = f_a(m) \times (f_a(n) \times a)$$
$$= f_a(m) \times f_a(n + 1)$$

and so $P(n + 1)$ is true and it follows by induction that

$$f_a(m + n) = f_a(m) \times f_a(n)$$

for all n.

We chose m arbitrarily, so the result is true for all m and all n.

(ii) Let $P(n)$ be the statement that $f_a(n) \times f_b(n) = f_{a \times b}(n)$. By definition,

$$f_a(1) \times f_b(1) = a \times b = f_{a \times b}(1),$$

so $P(1)$ is true. If $P(n)$ is true, then

$$
\begin{aligned}
f_a(n+1) \times f_b(n+1) &= (f_a(n) \times a) \times (f_b(n) \times b) \\
&= (f_a(n) \times f_b(n)) \times (a \times b) \\
&= f_{a \times b}(n) \times (a \times b) = f_{a \times b}(n+1),
\end{aligned}
$$

and so $P(n+1)$ is true. The result follows by induction.

(iii) Fix m and let $P(n)$ be the statement that $f_{u(m)}(n) = f_a(m \times n)$. From our definitions,

$$f_{u(m)}(1) = u(m) = f_a(m) = f_a(m \times 1),$$

so $P(1)$ is true. If $P(n)$ is true, then, using part (i),

$$
\begin{aligned}
f_{u(m)}(n+1) &= f_{u(m)}(n) \times u(m) = f_a(m \times n) \times f_a(m) \\
&= f_a((m \times n) + m) = f_a((m \times n) + (m \times 1)) \\
&= f_a(m \times (n+1)),
\end{aligned}
$$

so $P(n+1)$ is true. By induction $P(n)$ is true for all n.

Just as in (i), we observe that m was chosen arbitrarily, so the full result follows.   ∎

If we write $f_a(n) = a^n$, the conclusions of Theorem 4.2.4 take the familiar form

$$a^{m+n} = a^m \times a^n, \ (a \times b)^n = a^n \times b^n, \ (a^m)^n = a^{mn}.$$

These formulae are often called the *index laws*.

Exercise 4.2.5 *If $a \in \mathbb{Q}$ and $a \neq 0$, we define $a^0 = 1$. Check that the index laws continue to hold for the extended definition*[6].

Exercise 4.2.6 *If $a \in \mathbb{Q}$, $a \neq 0$ and n is an integer with $n \geq 0$, we define $a^{-n} = (a^{-1})^n$.*

(i) *Show, by induction, that $(a^n)^{-1} = a^{-n}$ for all integers $n \geq 1$. Deduce that $(a^n)^{-1} = a^{-n}$ for all integers.*

(ii) *Check that the index laws continue to hold for the extended definition.*

The method of proof of Theorem 4.2.4 will be echoed repeatedly when we consider the consequences of the Peano axioms in Chapter 6. If the reader

---

[6] Notice that we cannot extend *both* index laws $a^0 = 1$ and $0^n = 0$ to the case $a = 0, n = 0$. For this reason we leave 0^0 *undefined*. (Compare the footnote on page 39.)

thinks we have merely proved the obvious, I will not totally disagree, but I suspect that she did not find the index laws entirely obvious when she first met them at school.[7] More to the point, if the reader tries to explain exactly why these laws are obvious, she will find it hard to avoid the words 'and so on'. Mathematical induction makes the phrase 'and so on' precise.

Exercise 4.2.7

(i) *Show, by induction, that, if* $0 < a < b$, *and* n *is an integer with* $n \geq 1$, *then* $a^n < b^n$.

(ii) *Define* $n!$ *inductively*[8] *and prove that* $2^{n-1} \leq n! \leq n^n$ *for all integers* $n \geq 1$.

4.3 Applications

From now on, until Chapter 6, we shall take all the rules governing the integers as given. We will not only use these rules without comment, but we shall also use the usual conventions, writing

$$ab = a \times b, \ abc = a \times (b \times c), \ ab + cd = (a \times b) + (c \times d),$$

talking about $a - b = a + (-b)$ and so on.

Our object in this chapter is to show how the least member principle and its various equivalent forms and consequences can be used in practice. At least initially, our standpoint will not be that of farmers or tax collectors, but that of puzzle lovers, interested in numbers for their own sake.

We will need the following easy extension of the least element principle from $\mathbb{N}^+$ to $\mathbb{Z}$.

Lemma 4.3.1 *If* E *is a non-empty collection of integers bounded below (that is to say, there is a* $u \in \mathbb{Z}$ *with* $e \geq u$ *for all* $e \in E$*), then* E *has a least member.*

Proof. Let us write $a(e) = e + 1 - u$. Then $a(e) \geq 1$ for all $e \in E$ so, since the strictly positive integers may be identified with $\mathbb{N}^+$, the collection of $a(e)$ with $e \in E$ has a least member $a(e_0)$, say. Since

$$e = a(e) - 1 + u \geq a(e_0) - 1 + u = e_0,$$

we are done. ∎

---

[7] If you did find them obvious, generations of school teachers will sigh that you were an exceptional pupil.

[8] We write $n!$ for the factorial function. If the reader has not met this function before, she should look it up. It will be used in Chapter 5.

Exercise 4.3.2 *Show that a non-empty collection of integers bounded above has a greatest member.*

Exercise 4.3.3 *(This echoes Exercise 4.1.3.) We work in $\mathbb{Z}$. Suppose that $P(n)$ implies $P(n+1)$ for all $n \geq m$ and that $P(m)$ is true. Show that $P(n)$ is true for all $n \geq m$.*

Our first result is rather down to earth.

Lemma 4.3.4 [Long division] *If $n, m \in \mathbb{Z}$ and $m \geq 1$, then we can find $k, r \in \mathbb{Z}$ with $m > r \geq 0$ such that $n = km + r$.*

Proof. Consider the collection E of integers of the form $n - qm$ with $n - qm \geq 0$. Taking $q = 0$, if $n \geq 0$, and $q = n$, if $n < 0$, we see that E is non-empty. Since E is bounded below by 0, it has a least member $r = n - km$, say. If $r \geq m$, then $r' = n - (k+1)m \in E$ and $r > r'$, contradicting the minimality of r. Thus $m > r \geq 0$ and we are done. ∎

Definition 4.3.5 *We use the notation of Lemma 4.3.4. If $r = 0$, that is to say, $n = km$, we say that m divides k.*

Exercise 4.3.6 *Prove that long division gives a unique answer. More specifically show that if $m \in \mathbb{Z}$ with $m \geq 1$ and $k, k', r, r' \in \mathbb{Z}$ with $m > r \geq 0$, $m > r' \geq 0$ and $km + r = k'm + r'$, then $r = r'$ and $k = k'$.*

Our next result will also be familiar to most of my readers.

Lemma 4.3.7 *If u and v are natural numbers, then there is a largest natural number s with the property that s divides both u and v.*

Proof. Observe that the collection E of natural numbers which divide both u and v is non-empty since $1 \in E$. The collection E is bounded above because, if k divides u, then $k \leq u$. Thus E has a greatest member. ∎

Definition 4.3.8 *We use the notation of Lemma 4.3.7. The number s is called the* highest common factor *(or the* greatest common divisor*) of u and v. If $s = 1$, we say that u and v are* coprime.

Exercise 4.3.9

(i) *Find the highest common factor of 156 and 42.*
(ii) *Find the highest common factor of 107 748 and 69 126.*

[*Give part (ii) a good try, but do not worry if you cannot do it.*]

Lemma 4.3.7 tells us that a highest common factor exists, but gives no method other than brute force for finding it. The ancient Greeks discovered a very efficient way of finding highest common factors which we now call Euclid's algorithm.[9]

Euclid's algorithm *Given an ordered pair* (u, v) *with* u *and* v *integers and* $u \geq v \geq 1$, *either* v *divides* u *and we stop, recording* v *as our final answer, or* v *does not divide* u, *in which case*

$$u = kv + r$$

for some r *with* $v > r \geq 1$. *We replace the pair* (u, v) *with the pair* (u', v'), *where* $u' = v$ *and* $v' = r$, *and repeat our initial step.*

We claim that the process just described stops and when it stops the answer recorded is the highest common factor.

Exercise 4.3.10

(i) *Check that the algorithm works when we start with the pair* $(156, 42)$. *Choose your own pair of numbers and check that the algorithm works for them.*

(ii) *Check that the algorithm delivers an answer when we start with the pair* $(107\,748, 69\,126)$. *Check that the answer divides both integers in the pair. (Of course this does not show that we have the* highest *common factor.)*

(iii) *Write a computer program to implement Euclid's algorithm.*

Our proof that Euclid's algorithm works splits into two parts.

Lemma 4.3.11 *Euclid's algorithm terminates.*

Proof. We start with a pair (u, v) with $u \geq v \geq 1$. If v divides u, the process stops. If not, we form a new pair (u', v') with $u' = v$ and

$$u = kv + v'$$

for some k and some v' with $v > v' \geq 1$. Since the second term of each successive pair is strictly smaller than the second term of the previous pair, Exercise 4.1.13 tells us that the process must stop. ∎

Lemma 4.3.12 *Euclid's algorithm delivers the correct answer (that is to say, it delivers the highest common factor).*

---

[9] The Indian mathematicians, who either learned the method from the Greeks or rediscovered it for themselves, called it the 'pulveriser'.

Proof. Suppose that we apply one step of Euclid's algorithm to a pair (u, v) with $u \geq v \geq 1$. If v divides u, then v is indeed the highest common factor of u and v, so the algorithm has delivered the right answer.

If not, then the algorithm delivers a new pair (u', v') with $u' = v$ and

$$u = kv + v'$$

for some k and some v' with $v > v' \geq 1$. We look at d, the highest common factor of u and v and at d', the highest common factor of u' and v'.

Observe first that, since d is the highest common factor of u and v, we have $u = ad$, $v = bd$ for some natural numbers a and b, so $u' = bd$ and

$$v' = u - kv = ad - kbd = (a - kb)d.$$

Thus d divides u' and v' and, by definition, $d' \geq d$.

Next we observe that, since d' is the highest common factor of u' and v', we have $u' = a'd'$, $v' = b'd'$ for some natural numbers a' and b', so $v = u' = a'd'$ and

$$u = kv + v' = ka'd' + b'd' = (ka' + b')d'.$$

Thus d' divides u and v and, by definition, $d \leq d'$. Combining the results of this paragraph and its predecessor, we get $d = d'$.

Thus successive pairs in the Euclidean algorithm have the same highest common factor and, when the algorithm terminates, it will deliver the correct answer. ∎

We will make constant reference to the Euclidean algorithm during this part of the book, so it is particularly important that the reader is happy that she understands it.

Exercise 4.3.13 *Show that, if e divides u and v, then e divides the highest common factor d of u and v.*
[Hint. Show that, in the notation introduced for Euclid's algorithm, if e divides u and v, then e divides u' and v'. Exercise 4.3.19 (iv) gives another proof.]

The next two exercises ask the reader to apply the same sort of ideas as we used to verify the correctness of Euclid's algorithm to check the correctness of two other rather old (but much less important) algorithms.[10]

---

[10] When Adelard of Bath translated Al-Khwārizmī's treatise on Indian numerals into Latin, he Latinised the author's name as Algorismus. The methods used for calculating using the new system became known as *algorism* and this became *algorithm*: 'a set of rules for performing a calculation'.

Exercise 4.3.14 [Egyptian fractions]

(i) *Let p, q and k be strictly positive integers with $1/k > p/q$. Explain why there is a strictly positive integer $k' > k$ with*

$$\frac{1}{k'-1} > \frac{p}{q} \geq \frac{1}{k'}.$$

Show that, either $p/q = 1/k'$, or

$$\frac{p}{q} = \frac{1}{k'} + \frac{p'}{q'},$$

where p' and q' are strictly positive integers with $1/k' > p'/q'$ and $p > p'$. Explain how to calculate k' by looking at what happens when we divide q by p.

Explain why the result of the previous paragraph shows that every rational number x with $1 > x > 0$ can be written in the form

$$x = \frac{1}{k_1} + \frac{1}{k_2} + \ldots + \frac{1}{k_n}, \qquad \bigstar$$

with the k_j forming a strictly increasing sequence of strictly positive integers.

Sketch a computer program which, given strictly positive integers u and v, outputs the k_j in $\bigstar$ associated with the rational $x = u/v$.

(ii) *Apply the method of (i) to obtain*

$$\frac{4}{17} = \frac{1}{5} + \frac{1}{29} + \frac{1}{1233} + \frac{1}{3\,039\,345}.$$

Check that

$$\frac{4}{17} = \frac{1}{5} + \frac{1}{30} + \frac{1}{510},$$

so that although the method of (i) always produces an answer, it may not be the 'most economical'.

[*The method described in (i) appears in Fibonacci's* Liber Abaci *(the central text for the introduction of Indian and Arabic mathematics to Europe), though as one to be used only when all else fails. Today we would call it a 'greedy algorithm', since at each stage it takes 'the largest k' that will work'.*]

Exercise 4.3.15 [Russian peasant multiplication]

(i) *Suppose that we have an ordered triple (n, m, a) of integers with $n \geq m \geq 1$ and $a \geq 0$. If $m = 1$ we stop. Otherwise we proceed as follows.*

If m is even, we set $n' = 2n$, $m' = m/2$, $a' = a$. If m is odd, we set $n' = 2n$, $m' = (m-1)/2$ and $a' = a + n$. Show that the new triple (n', m', a') is a triple of integers with $n' \geq m' \geq 1$, $m > m'$ and $n'm' + a' = nm + a$.

Suppose that we start with the triple $(N, M, 0)$ where $N \geq M \geq 2$ and repeatedly apply the process described in the previous paragraph. Show that we must stop and, when we do stop, we will have a triple $(u, 1, w)$ with $MN = u + w$. Why does this justify the Russian peasant multiplication method described on page 13?

(ii) Use standard long multiplication with the binary notation to find 10110×1101. Compare what you have done with Russian peasant multiplication. Give a justification for Russian peasant multiplication based on the binary system.

The highest common factor takes centre stage in the next result.

Theorem 4.3.16 [Bézout's identity] *If u and v are non-zero integers, then we can find integers r and s such that*

$$ru + sv = d$$

where d is the highest common factor of $|u|$ and $|v|$.

Exercise 4.3.17

(i) *Suppose that u and v are non-zero integers and $|u|$ and $|v|$ have highest common factor d. Use Bézout's identity to show that we can write $n = au + bv$ for some integers a and b if and only if d divides n.*

(ii) *A team can score two types of goal in the game of phutball. The first is worth u points, the second v points. (Of course, u and v are strictly positive integers.) Let d be the highest common factor of u and v. Show that only scores of the form kd with k a non-negative integer are possible. Show, by means of an example, that some choices of scores of this form may not be possible. Show, however, that there always exists a K such that all scores of the form kd with $k \geq K$ are possible.*

Bézout's identity has a simple 'non-constructive' proof (that is to say a proof which establishes the existence of an object without showing how to calculate it[11]).

---

[11] Like most mathematicians and unlike most logicians for whom the term has precise technical meaning, I use the term 'non-constructive' rather loosely.

Non-constructive proof of Theorem 4.3.16. Consider the set E of strictly positive integers of the form $ru + sv$ with r and s integers. By taking $r = 1$ if $u > 0$, $r = -1$ if $u < 0$, $s = 1$ if $v > 0$, $s = -1$ if $v < 0$, we see that E is a non-empty subset of the strictly positive integers and so E has a least member $e = au + bv$.

We claim that e divides u. For if not, then, by long division,

$$u = te + s,$$

with $e > s > 0$ and

$$s = u - te = u - t(au + bv) = (1 - ta)u + (-(tb))v \in E,$$

contradicting the minimality of e. Similarly, e divides v and so e divides d.

However, any member of E and so, in particular, e, is divisible by d, so $d = e$. ∎

We can also use the Euclidean algorithm to give a constructive proof of Bézout's identity.

Constructive proof of Theorem 4.3.16. We may suppose that u and v are strictly positive integers with $u > v$. If we set $u = u_1$, $v = v_1$ and apply Euclid's algorithm, we obtain a sequence of pairs of strictly positive integers $(u_1, v_1), (u_2, v_2), \dots (u_n, v_n)$ together with strictly positive integers $a_1, a_2, \dots, a_n$ satisfying the equations

$$u_j = a_j v_j + v_{j+1}, \; u_{j+1} = v_j \qquad\qquad \bigstar_j$$

for $1 \le j \le n - 1$ and

$$u_n = a_n v_n.$$

We know that $v_n = d$.

We now 'reverse our steps'. By the last two sentences of the previous paragraph we know that

$$u_{n-1} = a_{n-1} v_{n-1} + d$$

and so

$$d = r_{n-1} u_{n-1} + s_{n-1} v_{n-1}$$

with $r_{n-1} = 1$ and $s_{n-1} = -a_{n-1}$. Now suppose that we have found integers r_{j+1} and s_{j+1}, such that

$$d = r_{j+1} u_{j+1} + s_{j+1} v_{j+1}$$

for some j with $n - 1 \geq j + 1 \geq 2$. Then the equations in $\bigstar_j$ give

$$d = r_{j+1}v_j + s_{j+1}(u_j - a_j v_j) = (r_{j+1} - s_{j+1}a_j)v_j + s_{j+1}u_j = r_j u_j + s_j v_j,$$

where

$$r_j = s_{j+1}, \quad s_j = r_{j+1} - s_{j+1}a_j.$$

Repeating this step for $j = n - 2$, $j = n - 2, \ldots, j = 1$ we obtain integers r_1 and s_1 such that

$$d = r_1 u_1 + s_1 v_1.$$

Taking $r = r_1, s = s_1, u = u_1$ and $v = v_1$ we have the desired result. ∎

Exercise 4.3.18 *The first sentence of the proof above asserts that we may take u and v strictly positive integers. What do we do if u and v are not both strictly positive?*

The method of proof we have just given provides an algorithm for computing r and s in Bézout's identity. We call it Bézout's algorithm.

Exercise 4.3.19

  (i) *Use Bézout's identity to show that, if u and v are coprime and ku = lv, then v divides k.*

 (ii) *Suppose u and v are coprime. Show that, if r and s are integers with*

$$ru + sv = 1,$$

then integers r' and s' also satisfy

$$r'u + s'v = 1$$

if and only if there exists an integer k such that $r - r' = kv$ and $s' - s = ku$.

(iii) *State and prove the corresponding result when u and v have highest common factor d.*

(iv) *Use Bézout's identity to prove the result of Exercise 4.3.13.*

Exercise 4.3.20 *It is often much easier to understand an algorithm after working through it in a few concrete cases. In my opinion, Bézout's algorithm is no exception. Apply the algorithm first to the pair 156 and 42. and then to the pair 107 748, 69 126 considered in Exercise 4.3.10 or to pairs of your own choosing.*

Exercise 4.3.21 *Write a computer program implementing Bézout's algorithm.*

Remark The reader may object that Euclid's algorithm predates Fermat and Pascal by two millennia.[12] Presumably the teachers of Egyptian scribes used the equivalent of 'and so on'. I ascribe the principle of the least member to Fermat and the principle of induction to Pascal because, so far as we know, they were the first to use these ideas systematically, explicitly and with a clear idea of their importance.

4.4 Prime Numbers

Hogben in his *Mathematics for the Million* [14] writes about the primes as follows.

> One class of numbers which attracted early interest is the *primes*. ...The recognition of this class of numbers was not a very useful discovery except in so far as it simplified finding square roots before modern methods were discovered.

However, we judge a rose or a poem not by its usefulness, but by the pleasure it gives. In this section we shall give results about the primes which go back to the ancient Greeks but whose proofs still give pleasure to mathematicians.

Definition 4.4.1 *We say that a natural number p is a* prime *if $p \neq 1$ and, whenever $p = uv$ with u and v natural numbers, either $u = 1$ or $v = 1$.*

Theorem 4.4.2 *Every natural number $m \neq 1$ can be written as the product of primes. (Here a single prime p is to be considered the product of primes.)*

Most mathematicians adopt the convention that 1 is the product of no primes and omit the condition $m \neq 1$.

Note The acute reader may observe that when we talk about the product of n numbers we are actually using an inductive definition and that the properties of such products need to be established by induction. I give such a definition and the associated proofs in Appendix A. However, we shall use these products only in this less formal chapter and the next and they form no part of the central arguments of this book. I suggest that the reader ignore the appendix or read it after Chapter 6.

Proof of Theorem 4.4.2. Let E be the collection of natural numbers which are neither 1 nor the product of primes. If E is empty we are done. If not, then, by the least member principle, E must have a least member e. We consider every

---

[12] If she is particularly well informed, she will further object that Euclid refers to least elements when proving some of the results discussed in this section and the next (see Book 7 of [8]).

prime as the product of primes, so e is not prime. Thus $e = uv$ with u and v natural numbers and u, $v \neq 1$. By the base element principle, $u > 1$ so, by the multiplication law for inequalities (Lemma 1.3.8), $e = u \times v > 1 \times v = v$ and, similarly, $e > u$. Since e is the least element of E, both u and v can be written as the product of primes and so e, being the product of u and v, is the product of primes. This contradiction shows that E must be empty. ∎

Moreover the 'prime decomposition' is unique in the following sense.

Theorem 4.4.3 [Uniqueness of prime factorisation] *If p_i $[1 \leq i \leq n]$ and q_j $[1 \leq j \leq m]$ are primes with*

$$p_1 \leq p_2 \leq \ldots \leq p_n, \ q_1 \leq q_2 \leq \ldots \leq q_m$$
$$and \ p_1 p_2 \ldots p_n = q_1 q_2 \ldots q_m,$$

then $n = m$ and $p_i = q_i$ $[1 \leq i \leq n]$.

Exercise 4.4.4 *Let S be the set of natural numbers which, when written in decimal, have a 1 in the unit place (so that $21 \in S$ and $8\,301 \in S$, but $23 \notin S$). Show that, if $a \in S$ and $b \in S$, then $ab \in S$. We say that $n \in S$ is* irreducible *if $n > 1$ and, whenever u, $v \in S$ satisfy $n = uv$, then $u = 1$ or $v = 1$. Show that every element n of S, with $n > 1$, can be written as the product of irreducibles. (An irreducible element is to be considered the product of irreducibles.) Show, however, that there exist distinct irreducibles a, b, c, d with $ab = cd$. [Hint: Consider $3 \times 7 \times 13 \times 17$.]*

Our proof of the uniqueness of prime factorisation depends on the following result which, in turn, depends on Bézout's identity.

Theorem 4.4.5 *If u and v are natural numbers and p is a prime which divides uv, then p must divide at least one of u and v.*

Proof. Write $uv = kp$. Suppose that p does not divide u. Since the only natural numbers dividing p are 1 and p itself, the highest common factor of u and p must be 1. Bézout's identity (Theorem 4.3.16) thus tells us that there exist integers a and b such that

$$au + bp = 1.$$

Simple algebra gives

$$v = (au + bp)v = a(uv) + (bv)p = a(kp) + (bv)p = (ak + bv)p,$$

so p divides v. ∎

Exercise 4.4.6 *If you have done Exercise 4.3.19 (i), check that Theorem 4.4.5 is a special case of that result.*

Exercise 4.4.7 *If u_1, u_2, ...u_n are natural numbers and p is a prime which divides $u_1 u_2 \ldots u_n$, show, by induction, that p must divide at least one of the u_j.*

Proof of Theorem 4.4.3. Suppose that there exists a natural number with two different prime factorisations. By the least element principle, there must exist a least such natural number; call it N. By definition, we can find primes p_i [$1 \le i \le n$] and q_j [$1 \le j \le m$] with

$$p_1 \le p_2 \le \ldots \le p_n, \; q_1 \le q_2 \le \ldots \le q_m$$
$$\text{and } N = p_1 p_2 \ldots p_n = q_1 q_2 \ldots q_m,$$

such that either $m \ne n$ or it is the case that $m = n$ and there exists a k with $1 \le k \le n$ such that $p_k \ne q_k$.

Without loss of generality, we may suppose $p_1 \le q_1$. We know that p_1 divides $N = q_1 q_2 \ldots q_m$ so, since p_1 is a prime, p_1 must divide at least one of the q_j. Since the q_j are primes, this means that p_1 must equal at least one of the q_j and so $q_1 = p_1$. Using the cancellation law for multiplication, we obtain

$$p_2 \le \ldots \le p_n, q_2 \le \ldots \le q_m$$
$$\text{and } p_2 \ldots p_n = q_2 \ldots q_m.$$

Thus the natural number $M = p_2 \ldots p_n$ has two different prime factorisations. Since $M < N$, this contradicts the statement that N is the least natural number with this property. ∎

Exercise 4.4.8 *Suppose that p_1, p_2, ...p_n are distinct primes and $r_1, r_2, \ldots r_n$, $s_1, s_2, \ldots s_n$ are integers with r_j, $s_j \ge 0$ [$1 \le j \le n$].*

(i) Show that, if

$$p_1^{r_1} p_2^{r_2} \ldots p_n^{r_n} = p_1^{s_1} p_2^{s_2} \ldots p_n^{s_n},$$

then $r_j = s_j$ for each j.

(ii) State and prove a necessary and sufficient condition on the r_j and s_j for $p_1^{r_1} p_2^{r_2} \ldots p_n^{r_n}$ and $p_1^{s_1} p_2^{s_2} \ldots p_n^{s_n}$ to be coprime (see Definition 4.3.8).

In his *A Mathematician's Apology* [13], Hardy illustrates his account with two

... 'simple' theorems, simple both in idea and in execution, but there is no doubt at all about their being theorems of the highest class. Each is as fresh and significant as when it was discovered – two thousand years have not written a wrinkle on either of them. Finally, both the statements and the proofs can be mastered in an hour by any intelligent reader ...

His first example is Euclid's proof of the existence of infinitely many primes.

Theorem 4.4.9 *There are infinitely many primes.*

Proof. Suppose that we could list the primes as $p_1, p_2, \ldots p_n$ say. Consider

$$N = (p_1 p_2 \ldots p_n) + 1.$$

We observe that if $1 \leq j \leq n$, then p_j does not divide N since it leaves remainder 1 when divided into N. However, we know that N factorises into primes, so there must be some prime not on our list.   ∎

Exercise 4.4.10 *Verify that* $(2 \times 3 \times 5 \times 7 \times 11 \times 13) + 1$ *is divisible by* 59. *Thus the* N *that appeared in our proof of Theorem 4.4.9 need not itself be a prime.*

His second theorem (associated with the School of Pythagoras) is very important for the study of numbers. (Since the Greeks thought geometrically, they interpreted the result in a different way from us.)

Theorem 4.4.11 *There is no* $x \in \mathbb{Q}$ *such that* $x^2 = 2$.

Proof. Suppose that $x \in \mathbb{Q}$ and $x^2 = 2$. By 'dividing top and bottom by 2 as many times as necessary' we may take $x = p/q$, where p and q are integers which are not both even.

Now $p^2/q^2 = x^2 = 2$, so $p^2 = 2q^2$. Since the square of an odd number is odd and $2q^2$ is even, p must be even; that is to say, $p = 2r$, where r is an integer. We now have $4r^2 = 2q^2$, so $2r^2 = q^2$ and, arguing as before, q is even. We started by assuming that p and q were not both even, so have arrived at a contradiction.   ∎

We now prove a natural generalisation, also found in Euclid's *Elements*.

Theorem 4.4.12 *Let* m *be a strictly positive integer. The equation*

$$x^2 = m$$

has no solution with $x \in \mathbb{Q}$ *unless* m *is the square of an integer.*

Proof. Suppose that x is a rational number with $x^2 = m$. We may suppose $x > 0$ and so we can write $x = u/v$ with u and v strictly positive integers. Since u and v are strictly positive integers, we can find distinct primes p_1, p_2, ... p_n and integers r_j, s_j, $t_j \geq 0$ $[1 \leq j \leq n]$ such that

$$m = p_1^{r_1} p_2^{r_2} \cdots p_n^{r_n}, \quad u = p_1^{s_1} p_2^{s_2} \cdots p_n^{s_n} \text{ and } v = p_1^{t_1} p_2^{t_2} \cdots p_n^{t_n}.$$

Since $m = x^2 = u^2/v^2$, we have $mv^2 = u^2$ and so

$$p_1^{r_1+2t_1} p_2^{r_2+2t_2} \cdots p_n^{r_n+2t_n} = p_1^{2s_1} p_2^{2s_2} \cdots p_n^{2s_n}.$$

By the uniqueness of factorisation, this gives

$$r_j + 2t_j = 2s_j$$

and so $r_j = 2(s_j - t_j)$ for each j. Since $r_j \geq 0$, it follows that $s_j - t_j \geq 0$ and writing

$$w = p_1^{s_1-t_1} p_2^{s_2-t_2} \cdots p_n^{s_n-t_n},$$

we have $m = w^2$ with w a strictly positive integer. ∎

Exercise 4.4.13 *Suppose that a and b are coprime strictly positive integers. Show that the equation $x^2 = a/b$ has a solution with $x \in \mathbb{Q}$ if and only if a and b are the squares of strictly positive integers.*

Exercise 4.4.14 *Magic tricks are rather dull without added pizzazz. Here the reader will have to add her own.*

Think of a letter belonging to the collection of sixteen written below.

$$A = \{a, b, c, d, e, f, g, h, i, j, k, l, m, n, o, p\}.$$

For each of the seven collections written below, write down 'yes' if the letter you have thought of belongs to the collection and 'no' otherwise. To make things more interesting, you may (but need not) lie once.

$$A_1 = \{b, c, f, g, i, l, m, p\}, A_2 = \{b, d, e, g, i, k, n, p\},$$
$$A_3 = \{b, d, f, h, j, l, n, p\}, A_4 = \{c, d, e, f, i, j, o, p\},$$
$$A_5 = \{c, d, g, h, k, l, o, p\}, A_6 = \{e, f, g, h, m, n, o, p\},$$
$$A_7 = \{i, j, k, l, m, n, o, p\}.$$

I now examine your answers. If you have answered yes an even number of times when asked about A_1, A_3, A_5 and A_7, I set $\eta_0 = 0$; otherwise I set $\eta_0 = 1$. If you have answered yes an even number of times when asked about A_2, A_3, A_6 and A_7, I set $\eta_1 = 0$; otherwise I set $\eta_1 = 1$. If you have answered *yes an even number of times when asked about* A_4, A_5, A_6 *and* A_7, *I set* $\eta_2 = 0$; *otherwise I set* $\eta_2 = 1$.

If $\eta_0 = \eta_1 = \eta_2 = 0$, *then you have not lied. Otherwise, you lied about* A_k, *where* $k = \eta_0 + 2\eta_1 + 4\eta_2$. *Moreover I can tell which letter you chose.*

How is the trick done?

[Do spend a little time thinking about this. An answer will be revealed in the next chapter.]

5

Modular Arithmetic

5.1 Finite Fields

There are many objects which share the algebraic properties of the rationals we gave earlier.

Definition 5.1.1 *A field* $(\mathbb{F}, +, \times)$ *is a set* $\mathbb{F}$ *together with two operations* $+$ *and* $\times$ *(with* $a + b \in \mathbb{F}$, $a \times b \in \mathbb{F}$ *whenever* a, $b \in \mathbb{F}$*) having the following properties for all* a, b, $c \in \mathbb{F}$.

(i) $a + b = b + a$. *(Commutative law of addition)*

(ii) $a + (b + c) = (a + b) + c$. *(Associative law of addition)*

(iii) *There exists an element* $0 \in \mathbb{F}$ *such that, whenever* $a \in \mathbb{F}$, $0 + a = a$. *(Additive zero)*

(iv) *For each* a *there exists an element* $-a \in \mathbb{F}$ *such that* $a + (-a) = 0$. *(Additive inverse)*

(v) $a \times b = b \times a$. *(Commutative law of multiplication)*

(vi) $a \times (b \times c) = (a \times b) \times c$. *(Associative law of multiplication)*

(vii) *There exists an element* $1 \in \mathbb{F}$ *such that, whenever* $a \in \mathbb{F}$, $1 \times a = a$. *(Multiplicative unit)*

(viii) *If* $a \neq 0$, *then there exists an* $a^{-1} \in \mathbb{F}$ *such that* $a \times a^{-1} = 1$. *(Multiplicative inverse)*

(ix) $a \times (b + c) = (a \times b) + (a \times c)$. *(Distributive law)*

We also demand $1 \neq 0$.

Exercise 5.1.2 *Compare Definition 5.1.1 with the statement of Theorem 3.2.7.*

Exercise 5.1.3 *State results corresponding to Exercise 3.2.11. Either recall the proofs from that exercise or reprove the results.*

Exercise 5.1.4 *Show that, if* $(\mathbb{F}, +, \times)$ *obeys all the numbered conditions of Definition 5.1.1, but we take* $0 = 1$, *then* $\mathbb{F}$ *consists of the single element* 0. *[See Exercise 3.2.14 (ii) if you need a hint.]*

Remark Nobody claims that 'the hard thing about chess is learning the rules'. When a child starts playing chess, the first few games they play may be devoted to learning the rules, but soon the rules are internalised as part of their tactical and later their strategic thinking. In the same way the reader will find that after working with various examples of fields the rules (or, as we say, axioms) for a field appear natural and easy to remember.

We have already written out long lists of rules for $\mathbb{N}$, $\mathbb{Q}^+$ and later on we will write down rules for objects like 'ordered fields' and 'skew fields', but they are all related. When a beginning cook learns a recipe for plum tart and then learns a recipe for apricot tart, the effort involved in learning the second recipe is much less and learning the second recipe may well make the general principles underlying the first recipe easier to understand.

From time to time, we will need two further definitions. The first makes precise the notion of isomorphism for fields. (We discussed the general notion of isomorphism in Section 3.4.)

Definition 5.1.5 *We say that the field* $(\mathbb{F}, +, \times)$ *is* isomorphic *to the field* $(\mathbb{G}, \oplus, \otimes)$ *if there exists a bijection* $f : \mathbb{F} \to \mathbb{G}$ *such that* $f(a + b) = f(a) \oplus f(b)$ *and* $f(a \times b) = f(a) \otimes f(b)$. *We say that* f *is an* isomorphism.

Exercise 5.1.6 *Suppose the conditions of Definition 5.1.5 hold. If* $\mathbb{F}$ *has additive zero* 0 *and multiplicative unit* 1, *identify* $f(0)$ *and* $f(1)$, *giving reasons for your answer.*

Definition 5.1.7 *Suppose that* $(\mathbb{F}, +, \times)$ *is a field. We say that a subset* $\mathbb{G}$ *of* $\mathbb{F}$ *is a* sub-field *of* $\mathbb{F}$ *if the following conditions hold.*

(i) If $a, b \in \mathbb{G}$, *then* $a + b$, $a \times b \in \mathbb{G}$.
(ii) If $a \in \mathbb{G}$, *then* $-a \in \mathbb{G}$ *and if, in addition,* $a \neq 0$, *then* $a^{-1} \in \mathbb{G}$.
(iii) $0, 1 \in \mathbb{G}$.

Exercise 5.1.8 *Suppose the conditions of Definition 5.1.7 hold. Check (this should not take you very long) that* $(\mathbb{G}, +, \times)$ *(where the operations are the natural inherited ones) is indeed a field.*

Exercise 5.1.9 *Check (again this should not take you very long) that we can replace* $(\mathbb{Q}, +, \times)$ *by any field* $(\mathbb{F}, +, \times)$ *in our discussion of powers* a^n *in Theorem 4.2.4, Exercise 4.2.5 and Exercise 4.2.6.*

We discuss one collection of fields, first explicitly recognised by Euler and Gauss.

Exercise 5.1.10 *Consider* $\mathbb{Z}$, *the collection of integers. Let n be an integer and write $r \sim_n s$ if $r - s = kn$ for some integer k. Show that $\sim_n$ is an equivalence relation on $\mathbb{Z}$ (see Definition 2.2.4).*

Identify the equivalence classes (see Definition 2.2.5) in the cases $n = 1$ and $n = 0$. Show that the equivalence classes are the same for $\sim_n$ and $\sim_m$ if and only if $m = n$ or $m = -n$.

From now on, we look at the collection of equivalence classes $\mathbb{Z}_n = \mathbb{Z}/\sim_n$ with $n \geq 2$. It turns out that we can define addition and multiplication on $\mathbb{Z}_n$ in a simple manner.

Lemma 5.1.11 *We work in $\mathbb{Z}$ and take $n \geq 2$.*

(i) If $u \sim_n u'$ and $v \sim_n v'$, then $u + v \sim_n u' + v'$.
(ii) If $u \sim_n u'$ and $v \sim_n v'$, then $uv \sim_n u'v'$.

Proof. We do part (ii). Since $u - u' = kn$ and $v - v' = ln$ for some integers l and k, we have

$$u'v' = (u - kn)(v - ln) = uv + (-kv - lu + kln)n,$$

so $u'v' - uv = (-kv - lu + kln)n$ and $uv \sim_n u'v'$. ∎

Exercise 5.1.12 *Prove part (i) of Lemma 5.1.11.*

We can thus make the unambiguous definitions

$$[a] \oplus [b] = [a + b], \ [a] \otimes [b] = [a \times b].$$

Theorem 5.1.13 *Let $n \geq 2$. The system $(\mathbb{Z}_n, \oplus, \otimes)$ satisfies all the conditions of Definition 5.1.1 with the possible exception of the existence of a multiplicative inverse.*

Proof. All the verifications are easy, using the corresponding results for $\mathbb{Z}$. We do four and leave the remainder to the reader.

(i) We have $[a] \oplus [b] = [a + b] = [b + a] = [b] \oplus [a]$.
(iii) We have $[0] \oplus [a] = [0 + a] = [a]$.
(iv) Observe that $[a] \oplus [-a] = [a - a] = [0]$.
(ix) We have

$$[a] \otimes ([b] \oplus [c]) = [a] \otimes [b + c] = [a(b + c)] = [(ab) + (ac)]$$

$$= [ab] \oplus [ac] = ([a] \otimes [b]) \oplus ([a] \otimes [c]). \qquad \blacksquare$$

Exercise 5.1.14 *Prove the remaining parts of Theorem 5.1.13.*

The system of arithmetic on $\mathbb{Z}_n$ is sometimes known as 'clock arithmetic'. If a twelve-hour clock shows u hours now, then after v hours, it will show w hours where (working in $\mathbb{Z}_{12}$) $[u] \oplus [v] = [w]$. However, whereas 'clock addition' is easy to understand,[1] I cannot see what 'clock multiplication' would correspond to.

Exercise 5.1.15 *Check that* $0 \sim_3 3$, *but* $2^0 \nsim_3 2^3$. *Thus we cannot define 'clock powers'* $[a]^{[b]}$.
[*We extend this remark in Exercise 5.2.3.*]

We shall use the index notation

$$[a]^u = [a^u],$$

but the reader should observe that, while $[a] \in \mathbb{Z}_n$, u is a strictly positive integer.

Exercise 5.1.16 *Do the one-line check that* $[a^1] = [a]$ *and* $[a^{n+1}] = [a^n] \times [a]$ *for* $n \in \mathbb{N}^+$. *There is thus no problem in writing* $[a]^n = [a^n]$.

Rather than talk about 'clock arithmetic' we shall refer to 'modular arithmetic' or 'arithmetic mod n'. In the standard notation, which we shall mainly (but not always) use, we write

$a \equiv c$	$\mod n$	to mean	$a \sim_n c$,
$a \not\equiv c$	$\mod n$	to mean	$a \nsim_n c$,
$a + b \equiv c$	$\mod n$	to mean	$[a] \oplus [b] = [c]$,
$ab \equiv c$	$\mod n$	to mean	$[a] \otimes [b] = [c]$.

We say that a and c are 'distinct modulo n' if $a \not\equiv c$.
Our first result shows that $(\mathbb{Z}_n, +, \times)$ may fail to be a field.

Exercise 5.1.17

(i) *Suppose that* $(\mathbb{F}, +, \times)$ *is a field. If* $u \neq 0$, *show, by multiplying by* u^{-1}, *that the equation* $uv = 0$ *implies* $v = 0$.

---

[1] Particularly for earlier generations familiar with clock faces rather than clock read-outs.

(ii) Suppose u, v are integers with u, v ≥ 2. If n = uv, show that

$$u \not\equiv 0 \mod n, \ v \not\equiv 0 \mod n, \ but \ uv \equiv 0 \mod n.$$

Conclude that $(\mathbb{Z}_n, +, \times)$ *is not a field.*

On the other hand, if p is a prime (and, by the previous result, *only* if p is prime), $(\mathbb{Z}_p, +, \times)$ is a field.

Theorem 5.1.18 *If p is a prime, $(\mathbb{Z}_p, +, \times)$ satisfies condition (viii) of Definition 5.1.1 and so, by Theorem 5.1.13, is a field.*

Proof. Suppose that $a \not\equiv 0 \mod p$. Then p does not divide a. Since p is prime (and so its only divisors are 1 and p), it follows that a and p have 1 as their highest common factor. Bézout's identity now tells us that there exist integers u and v with $up + va = 1$ and so with $va \equiv 1 \mod p$ (or, using our earlier language, $[v] \otimes [a] = [1]$). ∎

Exercise 5.1.19 *Suppose that $n \geq 2$, but n is not necessarily a prime. If a is coprime to n, show that there exists a strictly positive integer c such that $ac \equiv 1 \mod n$ and explain how to compute it using Bézout's algorithm.*

It is much easier to study an object if we have other similar objects to compare it with. In biology, the study of the genetics of fruit flies sheds light on the genetics of mankind.[2] In the same way, we might hope that the behaviour of $(\mathbb{Z}_p, +, \times)$ will shed light on the behaviour of $(\mathbb{Q}, +, \times)$.

5.2 Some Pretty Theorems

In this section we prove various results about $\mathbb{Z}_p$ when p is a prime.

Theorem 5.2.1 [Fermat's little theorem] *If $a \not\equiv 0 \mod p$, then $a^{p-1} \equiv 1 \mod p$.*

Proof. Since $\mathbb{Z}_p$ is a field (and, in particular, condition (viii) of Definition 5.1.1 holds), we know that if $[a] \neq [0]$, then $[a] \otimes [r] = [a] \otimes [s]$ implies $[r] = [s]$. Thus $[a]\otimes[1], [a]\otimes[2], \ldots, [a]\otimes[p-1]$ are distinct non-zero elements of $\mathbb{Z}_p$. Since $\mathbb{Z}_p$ has exactly $p - 1$ non-zero elements (namely $[1], [2], \ldots, [p - 1]$), it follows that $[a] \otimes [1], [a] \otimes [2], \ldots, [a] \otimes [p - 1]$ are the elements $[1], [2], \ldots, [p - 1]$ written down in some order. We thus have

---

[2] And, of course, vice versa.

$$([a] \otimes [1]) \otimes ([a] \otimes [2]) \otimes \cdots \otimes ([a] \otimes [p-1]) = [1] \otimes [2] \otimes \cdots \otimes [p-1].$$

Rearrangement gives

$$[a]^{p-1} \otimes ([1] \otimes [2] \otimes \cdots \otimes [p-1]) = [1] \otimes [2] \otimes \cdots \otimes [p-1],$$

and so [using condition (viii) again],

$$[a]^{p-1} = [1],$$

or, in more standard notation, $a^{p-1} \equiv 1 \mod p$.   ■

Exercise 5.2.2 *Let p be a prime. Show that $r^p \equiv r \mod p$ for all integers r.*

Exercise 5.2.3 *Generalise Exercise 5.1.15 by replacing 3 by p, where p is an odd prime. What happens when $p = 2$?*

Exercise 5.2.4 *Let $(\mathbb{F}, +, \times)$ be a field. If $\mathbb{F}$ has n elements, show that $x^{n-1} = 1$ whenever $x \neq 0$.*

Exercise 5.2.5 *If the reader knows the binomial theorem, here is another proof of Fermat's result. We suppose that p is a prime.*

  (i) *If r is an integer with $1 \leq r \leq p-1$, show that p does not divide $r!$ or $(p-r)!$. Deduce that $\binom{p}{r} \equiv 0 \mod p$.*
 (ii) *Deduce, by using the binomial expansion of $(k+1)^p$, that*

$$(k+1)^p \equiv k^p + 1 \mod p.$$

 (iii) *Use induction to prove that $r^p \equiv r$ for all natural numbers r and deduce Fermat's little theorem.*

We now study square roots in $\mathbb{Z}_p$.

Lemma 5.2.6 *Let p be an odd prime. The equation*

$$x^2 \equiv a \mod p \qquad\qquad ★$$

(with $0 \leq x \leq p-1$) has exactly one solution modulo p for x if $a \equiv 0$. Otherwise, it has either 2 or 0.

Proof. We leave the second sentence of our statement as an exercise.

If ★ has a solution $b \not\equiv 0$, then $a \equiv b^2$. Since $\mathbb{Z}_p$ is a field, we know that, if $x - b \not\equiv 0$ and $x + b \not\equiv 0$, we have $(x-b)(x+b) \not\equiv 0 \mod p$ so

$$x^2 - a \equiv x^2 - b^2 = (x-b)(x+b) \not\equiv 0.$$

Thus the only solutions of ★ are $x \equiv b$ and $x \equiv -b$.

If $b \equiv -b$, then $2b \equiv 0 \mod p$. Since $2 \not\equiv 0$ it follows that $b \equiv 0$. Thus, if $a \not\equiv 0$, ★ has 2 or 0 distinct solutions modulo p. ∎

Exercise 5.2.7 *Prove the second sentence in the statement of Lemma 5.2.6.*

Exercise 5.2.8 *We use the notation of Lemma 5.2.6 and let p be an odd prime. Show that there are exactly $(p+1)/2$ values of a (modulo p) such that ★ has a solution.*

Exercise 5.2.9 *What can we say about solutions of ★ if $p = 2$?*
Mathematicians are very interested in square roots of -1. If we deal with $\mathbb{Z}_p$, the matter is rather simple. We use the following result.

Lemma 5.2.10 [**Wilson's theorem**] *If p is an odd prime, then $(p - 1)! \equiv -1 \mod p$.*

Proof. If $u \not\equiv 1, -1, 0$, then (since Lemma 5.2.6 tells us that $u^2 \not\equiv 1$) we know that $u \not\equiv u^{-1}$. Thus we can pair off those elements $u \not\equiv 1, -1, 0$ of $\mathbb{Z}_p$ as $u \leftrightarrow v$ with $uv = 1$. Now $2 \times 3 \times 4 \ldots \times (p - 2)$ is the product of these pairs, so

$$2 \times 3 \times 4 \ldots \times (p - 2) \equiv 1 \mod p$$

and

$$(p - 1)! \equiv 1 \times 2 \times 3 \times 4 \ldots \times (p - 2) \times (p - 1)$$
$$\equiv 1 \times \left(2 \times 3 \times 4 \ldots \times (p - 2)\right) \times (p - 1)$$
$$\equiv 1 \times 1 \times (-1) \equiv -1 \mod p,$$

as stated. ∎

Exercise 5.2.11 *What happens if $p = 2$?*

Lemma 5.2.12

(i) *If p is a prime with $p = 4n + 3$ for some non-negative integer n, then the equation*

$$x^2 \equiv -1 \mod p$$

has no solution.

(ii) *If p is a prime with $p = 4n + 1$ for some non-negative integer n, then the equation*

$$x^2 \equiv -1 \mod p$$

has exactly two solutions modulo p.

(iii) If p = 2, the equation

$$x^2 \equiv -1 \mod p$$

has exactly one solution modulo p.

Proof. (i) Suppose, if possible, that $x^2 \equiv -1 \mod p$. Then, by Fermat's little theorem,

$$-1 \equiv (-1)^{2n+1} \equiv (x^2)^{2n+1} \equiv x^{4n+2} \equiv x^{p-1} \equiv 1 \mod p,$$

which is absurd.

(ii) Set $x \equiv 1 \times 2 \times \cdots \times 2n$. Then, since

$$2n + r \equiv -\big(p - (2n+r)\big) \equiv -(2n + 1 - r) \mod p,$$

it follows that

$$(2n + 1) \times (2n + 2) \times \cdots \times 4n \equiv \big(-(2n)\big) \times \big(-(2n-1)\big)$$
$$\times \cdots \times (-1)$$
$$\equiv (-1)^{2n} \times (2n) \times (2n-1)$$
$$\times \cdots \times 1 \equiv x.$$

Thus, using Wilson's theorem,

$$x^2 \equiv \big(1 \times 2 \times \cdots \times (2n)\big) \times \big((2n+1) \times (2n+2) \times \cdots \times (4n)\big) \equiv (4n)! \equiv -1.$$

Lemma 5.2.6 tells us that there are exactly two solutions, x and $-x$.

(iii) By inspection, 1 is the unique solution. ∎

Part (iii) of the next lemma will play a very important role in this chapter.

Lemma 5.2.13

(i) *If p is an odd prime and $a \not\equiv 0$, then $a^{(p-1)/2} \equiv 1$ or $a^{(p-1)/2} \equiv -1$ mod p.*

(ii) *If p is a prime with $p = 4n + 3$ for some non-negative integer n, and $a \not\equiv 0$, then $a^{(p-1)/2} \equiv 1$.*

(iii) *If p is a prime with $p = 4n + 3$ for some non-negative integer n, then, if $a \equiv b^2 \mod p$, we have $b \equiv a^{(p+1)/4}$ or $b \equiv -a^{(p+1)/4} \mod p$.*

Proof. (i) Let $c \equiv a^{(p-1)/2}$. Then, by Fermat's little theorem, $c^2 \equiv a^{p-1} \equiv 1$. Since the equation $x^2 \equiv 1$ has at most two roots, we know that $c \equiv 1$ or $c \equiv -1$ and we are done.

(ii) We use part (i) together with Lemma 5.2.12 (i).

(iii) If $b \equiv 0$, the result is trivial. Otherwise, we know that a has exactly two square roots and part (ii) gives

$$(-a^{(p+1)/4})^2 \equiv (a^{(p+1)/4})^2 \equiv a^{(p+1)/2} \equiv a^{(p-1)/2} \times a \equiv a,$$

as stated. ∎

Thus it is easy to find square roots in $\mathbb{Z}_p$ with p a prime if $p \equiv -1 \mod 4$. (The problem is harder when $p \equiv 1 \mod 4$.)

5.3 A New Use for Old Numbers

At school we learn to count and to write in different classes. Letters make up words which describe things, whereas numbers count things. In this section we use numbers to make up words.

The first extensive system for sending messages over long distances faster than a man on a horse was the aerial telegraph of Chappe, developed during the French Revolution, initially for military use. It used semaphore towers with different positions of the semaphore arms standing for different letters or words. Each tower would repeat the signal observed at one of its neighbours and this repeated signal would, in turn, be observed and repeated by the next tower along the route.[3]

The system was very successful, but now lives only in literature, where it occurs in Napoleonic sea stories, *The Count of Monte Cristo* and, more recently, as the 'clacks towers' in the novels of Terry Pratchett.

The aerial telegraph was rendered obsolete by the electric telegraph. The successful Morse system now required the message to be translated at one end into a series of dots, dashes and pauses and then translated back at the other

---

[3] If you'll only just promise you'll none of you laugh
I'll be after explaining the French Telegraphe!
A machine that's endowed with such wonderful pow'r
It writes, reads and sends news 50 miles in an hour.
Then there's watchwords, a spy-glass, an index on hand
And many things more none of us understand,
But which, like the nose on your face, will be clear
When we have as usual improved on them here.
Adieu, penny posts! mails and coaches, adieu!
Your Occupation's gone, 'tis all over wid you.
In your place telegraphs on our houses we'll see
To tell time, conduct lightning, dry shirts and send news.
Charles Dibden, Great News or a Trip to the Antipodes, *play produced in 1794*

end. Thus M corresponds to $--$, O to $---$, R to $\bullet - \bullet$ and so on. If we write $\star$ for pause, then *MORSE CODE* becomes

$$- -\star - - -\star\bullet - \bullet\star\bullet\bullet\bullet\star\bullet\star\star - \bullet - \bullet\star - - - -\star - \bullet\bullet\star\bullet\star.$$

Exercise 5.3.1 *Write* DO MORE *in Morse code.*

Since the cost of transmitting messages depended on their length, Morse's system gave shorter codes for more frequent letters like E. Although mistakes in transmission must have occurred quite often, the fact that the translation back from Morse into English was done by a human being meant that grabled messoges could usually be ungarbled.

As translation into and from code became mechanised, it was realised that the new systems would operate more smoothly if pauses were dispensed with and messages written using only $\bullet$ and $-$. It was natural to think of $\bullet$ as 0 and $-$ as 1, so

$$\bullet - \bullet\bullet - - - \bullet \text{ became } 01001110.$$

Exercise 5.3.2 *The object of this exercise is to show that we need only two symbols, 0 and 1, say, to transmit messages. We use 'words' consisting of five symbols. The word* 00000 *corresponds to* $\star$ *and the kth letter of the alphabet corresponds to the 'word'* $\epsilon_0\epsilon_1\epsilon_2\epsilon_3\epsilon_4$, *where each* ϵ_j *is 0 or 1 and*

$$k = \epsilon_0 2^4 + \epsilon_1 2^3 + \epsilon_2 2^2 + \epsilon_3 2 + \epsilon_4.$$

(i) *Check that L corresponds to* 01100.

(ii) *Translate the following English message into code: ALL$\star$WELL$\star\star$. You may find this table[4] useful.*

	000	001	010	011	100	101	110	111
00	$\star$	A	B	C	D	E	F	G
01	H	I	J	K	L	M	N	O
10	P	Q	R	S	T	U	V	W
11	X	Y	Z					

(iii) *Translate the following message into English.*

10111001010110001100000000010001111011110001010000000000

In the 1940s many of the systems used 'eight-hole paper-tape' to record the message in its two symbol form. Essentially, the message was split into 'words'

---

[4] Tables like these are called font tables. The word font comes from the French 'fondre', meaning to melt or to cast. It was adopted by printers to mean the collection of cast metal type used to type set a page. The 'font tables' used in producing this book show the correspondence between the various symbols and their binary representations.

consisting of eight 'letters' 0 or 1. The letter 1 was represented by a hole in the tape and the letter 0 by its absence. Such a tape could be read very quickly by an optical reader and was used in early computer installations.

Without a human being to detect garbled messages, it became very important to have some automatic means of detecting errors. This was done by using only words $\zeta_1 \zeta_2 \ldots \zeta_8$ (where each ζ_j is either 0 or 1) such that

$$\zeta_1 + \zeta_2 + \ldots + \zeta_8 \equiv 0 \quad \text{mod } 2. \qquad \bigstar$$

Since ζ_j could be chosen freely for $1 \le j \le 7$ and then

$$\zeta_8 \equiv \zeta_1 + \zeta_2 + \ldots + \zeta_7 \quad \text{mod } 2,$$

the entry ζ_8 was usually referred to as the 'check digit' and the system itself as a *check digit code*.

Exercise 5.3.3

(i) *Check the arithmetic connecting the last two displayed formulae. Does anything, apart from human nature, prevent us from considering ζ_3 as the check digit?*

(ii) *(A very old joke.) Is it possible for 7 people to share 20 lumps of sugar in such a way that each gets an odd number of lumps in their tea?*

With this system the machine could check each eight-letter word that it received and signal an error if $\bigstar$ was not satisfied. Of course, $\bigstar$ will fail only if there are an odd number of mistakes, but the system was used only when errors were rare and it was very unlikely that more than one error would occur in one line (that is to say, one word).

Exercise 5.3.4 *Suppose that ζ_j, $\zeta_j' \in \mathbb{Z}_2$ for $1 \le j \le 8$. Check that, if*

$$\zeta_1 + \zeta_2 + \cdots + \zeta_8 \equiv 0 \quad \text{mod } 2,$$

then

$$\zeta_1' + \zeta_2' + \cdots + \zeta_8' \equiv 0 \quad \text{mod } 2$$

if and only if $\zeta_j \ne \zeta_j'$ for an even number of values of j.

Exercise 5.3.5 *If you look at the inner title page of almost any book published between 1970 and 2006, you will find its International Standard Book Number (ISBN). The ISBN uses single digits selected from 0, 1, ..., 8, 9 and X* *representing* 10. *Each ISBN consists of nine such digits* $a_1, a_2, \ldots, a_9$ *followed by a single check digit* a_{10} *chosen so that*

$$10a_1 + 9a_2 + \cdots + 2a_9 + a_{10} \equiv 0 \quad \text{mod } 11. \qquad (*)$$

  (i) *Find a couple of books published in the given period and check that* $(*)$ *holds for their ISBNs.*[5]

 (ii) *Show that* $(*)$ *will not hold if you make a mistake in writing down one digit of an ISBN.*

(iii) *Show that* $(*)$ *may hold if you make two errors.*

 (iv) *Show that* $(*)$ *will not work if you interchange two distinct adjacent digits (a transposition error).*

  (v) *Does (iv) remain true if we replace 'adjacent' by 'different'?*

    Errors of the sort (ii) and (iv) are the most common in typing. In communication between publishers and booksellers, both sides are anxious that errors should be detected, but would prefer the other side to query errors rather than to guess what the error might have been.

 (vi) *For books published after January 2007, the appropriate ISBN is a thirteen-digit number* $x_1 x_2 \ldots x_{13}$ *with each digit selected from* $0, 1, \ldots,$ 8, 9 *and the check digit* x_{13} *computed by using the formula*

$$x_{13} \equiv -(x_1 + 3x_2 + x_3 + 3x_4 + \cdots + x_{11} + 3x_{12}) \quad \text{mod } 10.$$

Show that we can detect single errors. Give an example to show that we cannot detect all transpositions.

As a fairly junior researcher, Hamming had access to an early electronic computer, but was low down in the priority list of users. He would submit his programs encoded on paper-tape to run over the weekend, but often he would have his tape returned on Monday because the machine had detected an error in the tape. 'If the machine can detect an error', he asked himself, 'why can the machine not correct it?' He now came up with the following scheme.

We work in $\mathbb{Z}_2$ and use words of length 7 (so that, in effect, we work with seven-hole tape). The words $c_1 c_2 c_3 \ldots c_7$ with $c_j \in \mathbb{Z}_2$ are chosen to satisfy the three conditions

$$c_1 + c_3 + c_5 + c_7 \equiv 0 \quad \text{mod } 2 \qquad (0)$$

$$c_2 + c_3 + c_6 + c_7 \equiv 0 \quad \text{mod } 2 \qquad (1)$$

$$c_4 + c_5 + c_6 + c_7 \equiv 0 \quad \text{mod } 2. \qquad (2)$$

---

[5] In fact, X is used only in the check digit place.

By inspection, we may choose c_3, c_5, c_6 and c_7 freely and then c_1, c_2 and c_4 are completely determined.

Exercise 5.3.6 *Write down formulae which give c_1, c_2 and c_4 in terms of the other c_j. Explain why we have exactly 2^4 possible words.*

The way in which Hamming chose his conditions may become clearer if we use a binary expansion. Observe that (working in $\mathbb{Z}$) each integer r with $1 \le r \le 7$ can be written uniquely as

$$r = \epsilon_0(r) + \epsilon_1(r)2 + \epsilon_2(r)2^2,$$

with $\epsilon_j(r)$ taking the values 0 or 1. Hamming's three conditions thus become

$$\text{(The sum of those } c_r \text{ with } \epsilon_j(r) = 1) \equiv 0 \quad \mod 2 \qquad (j)$$

or more briefly

$$\sum_{\epsilon_j(r)=1} c_r \equiv 0 \quad \mod 2 \qquad (j)$$

for $j = 0$, $j = 1$, $j = 2$.

Lemma 5.3.7 *Suppose that we have c_r satisfying Hamming's conditions and some k with $1 \le k \le 7$.*
Consider c_r' with $c_r' = c_r$ for $r \ne k$ and $c_k \ne c_k'$. Then

$$\sum_{\epsilon_j(r)=1} c_r' \equiv 0 \quad \mod 2$$

if $\epsilon_j(k) = 0$ and

$$\sum_{\epsilon_j(r)=1} c_r' \equiv 1 \quad \mod 2$$

if $\epsilon_j(k) = 1$.

Proof. Observe that, if $\epsilon_j(k) = 0$, then $r \ne k$ if $\epsilon_j(r) = 1$, so

$$\sum_{\epsilon_j(r)=1} c_r' \equiv \sum_{\epsilon_j(r)=1} c_r \equiv 0 \quad \mod 2.$$

On the other hand, if $\epsilon_j(k) = 1$,

$$\sum_{\epsilon_j(r)=1} c_r' \equiv (c_k' - c_k) + \sum_{\epsilon_j(r)=1} c_r \equiv 1 + 0 \equiv 1 \quad \mod 2. \qquad \blacksquare$$

Suppose that we know that the 'sent word' $c_0c_1 \ldots c_7$ satisfies Hamming's conditions and we know that the 'received word' $c_0'c_1' \ldots c_7'$ differs from the sent word in at most one place. Then Lemma 5.3.7 allows us to recover the sent word as follows.

If the received word satisfies all the Hamming conditions, then it must be the sent word. Otherwise, we proceed as follows. If the received word satisfies the jth Hamming condition, that is to say,

$$\sum_{\epsilon_j(r)=1} c_r' \equiv 0 \mod 2, \tag{j}$$

write $\eta_j = 0$. If the received word fails to satisfy the jth Hamming condition, that is to say,

$$\sum_{\epsilon_j(r)=1} c_r' \equiv 1 \mod 2, \tag{j'}$$

write $\eta_j = 1$. Then $k = \eta_0 + 2\eta_1 + 2^2\eta_2$ and $c_j = c_j'$ for $j \neq k$, $c_k \equiv c_k' + 1$ mod 2.

It is easy to see that Hamming's idea works only for systems where errors are fairly rare.

Exercise 5.3.8 *If the received word differs from the sent word in two or more places, show that the Hamming scheme will always deliver the wrong answer for the sent word.*

Exercise 5.3.9 *Check that, if we replace letters by numbers, as shown below, we obtain (an isomorphic copy of) the trick in Exercise 4.4.14.*

$$A = \{0, 7, 25, 30, 42, 45, 51, 52, 131, 152, 162, 165, 193, 198, 240, 255\},$$

$$A_1 = \{7, 25, 45, 51, 131, 165, 193, 255\}, \quad A_2 = \{7, 30, 42, 51, 131, 162, 198, 255\},$$

$$A_3 = \{7, 30, 45, 52, 152, 165, 198, 255\}, \quad A_4 = \{25, 30, 42, 45, 131, 152, 240, 255\},$$

$$A_5 = \{25, 30, 51, 52, 162, 165, 240, 255\}, \quad A_6 = \{42, 45, 51, 52, 193, 198, 240, 255\},$$

$$A_7 = \{131, 152, 162, 165, 193, 198, 240, 255\}.$$

Explain why the trick works and how to find the number that you have chosen. (If you have already done Exercise 4.4.14, think how your answer relates to this section.)

Exercise 5.3.10 *(This exercise needs a smidgen of probability.)*

(i) *Suppose that we use eight-hole tape with the standard check digit code (see page 93) and the probability that an error occurs at a particular place on the tape (that is to say, a hole occurs where it should not or fails to occur where it should) is* 10^{-4} *independently of what happens at any other place. A program requires about 10 000 lines of tape (each line containing eight places) using the paper-tape code. Show (by using a calculator or, if you know it, the Poisson approximation) that the probability that the tape will be accepted as error free by the decoder is less than 0.04%.*

Suppose now that we use the Hamming scheme (making no use of the last place in each line). Explain why the program requires about 17 500 lines of tape, but the probability that any particular line will be correctly decoded is about $1 - (21 \times 10^{-8})$ *and the probability that the entire program will be correctly decoded is better than 99.6%.*

(ii) Suppose the error rate is 10^{-1} *rather than* 10^{-4}, *but the other conditions remain the same as in (i). Use Exercise 5.3.8 to show that the Hamming scheme is useless.*

Hamming's scheme is easy to implement. It took a little time for his company to realise what he had done,[6] but it was soon trying to patent the idea.

There are entire university courses devoted to the design of codes which work when the error rate is high. The next exercise suggests what we might do if the error rate is very low.

Exercise 5.3.11 *Construct a Hamming-type code for 'fifteen-hole tape'. Explain why it will be useful only when the error rate is very low. By considering the cost of transmitting a single symbol (that is to say, 0 or 1), explain why the new system is cheaper to use in this case.*

It is much easier to appreciate Newton's first law that 'a body will remain in a state of uniform motion unless acted on by some external force' if you drive a sports car than if you drive a horse and cart. In the same way, it is much easier for us to understand that a book, a symphony or a picture may be translated into a series of zeros and ones, that is to say, a number, than it would have been for our ancestors, since we are surrounded by machines which do just that. However, Leibniz, who invented the binary notation for representing integers, might not have been surprised since he thought that his system illustrated

> ... the creation of all things out of nothing through God's omnipotence, it might be said that nothing is a better analogy to, or even demonstration of such creation than the origin of numbers as here represented, using only unity and zero or nothing.

---

[6] Experienced engineers came away from working demonstrations muttering 'I still don't believe it'.

And it would be difficult to find a better illustration of this secret in nature or philosophy ...

Letter to the Duke of Brunswick, January 1697

5.4 More Modular Arithmetic

We have seen that $\mathbb{Z}_n$ is a field if and only if n is a prime. However, we can still do quite a lot of arithmetic when n is not a prime.

Our main tool is a result called the Chinese remainder theorem which can been traced back to a third-century Chinese mathematician named Sun Zi, who stated the following problem and gave its solution,[7]

> Now there are an unknown number of things. If we count by threes, there is a remainder 2; if we count by fives, there is a remainder 3; if we count by sevens, there is a remainder 2. Find the number of things.
>
> *In* The Mathematical Classic of Sun Zi

In our notation, Sun Zi's problem asks for the solution of

$$x \equiv 2 \quad \mathrm{mod}\ 3$$
$$x \equiv 3 \quad \mathrm{mod}\ 5$$
$$x \equiv 2 \quad \mathrm{mod}\ 7.$$

Let us see if we can find a general method for the simplest version of the problem, when we have *two* equations

$$x \equiv a_1 \quad \mathrm{mod}\ n_1$$
$$x \equiv a_2 \quad \mathrm{mod}\ n_2$$

with n_1 and n_2 coprime.

Reflection suggests that we should start by trying to solve our problem in the case

$$x_1 \equiv 1 \quad \mathrm{mod}\ n_1$$
$$x_1 \equiv 0 \quad \mathrm{mod}\ n_2.$$

We may then have the happy thought that (since n_1 and n_2 are coprime) Bézout's algorithm gives us a quick procedure for finding integers u_1 and u_2 such that

$$u_1 n_1 + u_2 n_2 = 1.$$

---

[7] In *A Mathematician's Apology* [13], Hardy refers to the anonymity of mathematical fame. All we know about Sun Zi is that he wrote the book in which this problem occurs.

Then, if we set $x_1 = u_2 n_2$,

$$x_1 \equiv u_2 n_2 \equiv 1 - u_1 n_1 \equiv 1 \quad \text{mod } n_1$$
$$x_1 \equiv u_2 n_2 \equiv 0 \quad \text{mod } n_2.$$

Exercise 5.4.1 *Find a solution to the pair of equations*

$$x_2 \equiv 0 \quad \text{mod } n_1$$
$$x_2 \equiv 1 \quad \text{mod } n_2.$$

More generally, we see that, if

$$x = a_1 u_2 n_2 + a_2 u_1 n_1,$$

then

$$x \equiv a_1 \quad \text{mod } n_1$$
$$x \equiv a_2 \quad \text{mod } n_2$$

and we have found a solution to our problem.

Lemma 5.4.2 [Chinese remainder theorem for two equations] *If n_1 and n_2 are coprime integers, the system of equations*

$$x \equiv a_1 \quad \text{mod } n_1$$
$$x \equiv a_2 \quad \text{mod } n_2$$

has exactly one solution $x \equiv x_0$ modulo $n_1 n_2$.

Proof. We have already seen that the system has a solution y_0. Long division shows that there exists an m such that, setting $x_0 = y_0 - m n_1 n_2$, we have $0 \leq x_0 \leq n_1 n_2 - 1$. An easy calculation shows that x_0 solves our equations, as does any $x = x_0 + k n_1 n_2$ with k an integer.

On the other hand, if

$$y \equiv a_1 \quad \text{mod } n_1$$
$$y \equiv a_2 \quad \text{mod } n_2,$$

then

$$y - x_0 \equiv 0 \quad \text{mod } n_1$$
$$y - x_0 \equiv 0 \quad \text{mod } n_2.$$

Since the coprime integers n_1 and n_2 divide $y - x_0$, we know (using the uniqueness of factorisation or Bézout's identity) that $n_1 n_2$ divides $y - x_0$. Thus $y - x_0 = k n_1 n_2$, with k an integer and so $y \equiv x_0 \text{ mod } n_1 n_2$. ∎

We note the useful special case of Lemma 5.4.2.

Lemma 5.4.3 *If n_1 and n_2 are coprime integers, then*

$$x \equiv a \quad \mod n_1$$
$$x \equiv a \quad \mod n_2$$

if and only if $x \equiv a \mod n_1 n_2$.

We shall only use the Chinese remainder theorem two modular equations, but our method is easily extended. Suppose that n_1, n_2, n_3 are strictly positive integers with each pair n_i, n_j $[i \neq j]$ coprime. We can solve the system

$$x \equiv a_1 \quad \mod n_1$$
$$x \equiv a_2 \quad \mod n_2$$
$$x \equiv a_3 \quad \mod n_3$$

as follows. We use the method for dealing with two equations to find y such that

$$y \equiv a_1 \quad \mod n_1$$
$$y \equiv a_2 \quad \mod n_2.$$

We now observe that $n_1 n_2$ and n_3 are coprime so we can solve

$$x \equiv y \quad \mod n_1 n_2$$
$$x \equiv a_3 \quad \mod n_3.$$

It is easy to check that x is a solution to our initial problem.

Exercise 5.4.4 *Solve Sun Zi's problem.*

The remainder of this section is given over to exercises which the reader may find interesting, but will not be used later.

Exercise 5.4.5 [General Chinese remainder theorem] *Extend Lemma 5.4.2 to the system of modular equations*

$$x \equiv a_j \quad \mod n_j \quad [1 \leq j \leq m]$$

(with each pair n_i, n_j coprime $[i \neq j]$) by induction on m.

Exercise 5.4.6 *If n is a strictly positive integer with $n \geq 5$ which is not a prime, show that $(n-1)! \equiv 0 \mod n$.*
What happens if $1 \leq n \leq 4$?
[Compare Wilson's theorem (Lemma 5.2.10).]

Exercise 5.4.7 *Suppose that n_1 and n_2 are integers with n_1, $n_2 \geq 2$ having highest common factor d.*

(i) Show that, if the system of equations

$$x \equiv a_1 \quad \mathrm{mod}\ n_1$$
$$x \equiv a_2 \quad \mathrm{mod}\ n_2$$

has a solution, then $a_1 - a_2 \equiv 0 \mod d$.

(ii) Use Bézout's theorem to show that there exists an integer y such that

$$y \equiv 0 \quad \mathrm{mod}\ n_1$$
$$y \equiv d \quad \mathrm{mod}\ n_2.$$

(iii) By considering $x = a_1 + kd$, or otherwise, show that, if $a_1 - a_2 \equiv 0$ mod d, then the system of equations given in (i) has a solution.

5.5 Problems of Equal Difficulty

In this section we shall be interested in $(\mathbb{Z}_{pq}, +, \times)$, where p and q are distinct odd primes. More specifically, we shall be interested in square roots for this system.

Lemma 5.5.1 *Let p and q be distinct odd primes.*

(i) The equation $x^2 \equiv 1 \mod pq$ has four distinct roots 1, -1 and two further roots[8] ϵ and $-\epsilon$.

(ii) If $a \not\equiv 0 \mod p$ and $a \not\equiv 0 \mod q$, then the equation $x^2 \equiv a^2 \mod pq$ has four distinct roots a, $-a$, ϵa and $-\epsilon a$.

Proof. (i) By Lemma 5.2.6, the equation

$$x^2 \equiv 1 \quad \mathrm{mod}\ p \qquad\qquad \bigstar$$

has solutions $x \equiv 1$ and $x \equiv -1$ and Lemma 5.2.6 tells us that these are the only solutions. The same result holds with p replaced by q, so the Chinese remainder theorem (Lemma 5.4.2) tells us that the equation $x^2 \equiv 1 \mod pq$ has four distinct roots given by the four different sets of modular equations

---

[8] If the reader finds the idea of 'four square roots' a bit odd, she should note that still stranger things happen in other algebraic systems [see Exercise 11.1.10 (iii)].

$$\begin{cases} x \equiv 1 \mod p \\ x \equiv 1 \mod q \end{cases} \quad \begin{cases} x \equiv -1 \mod p \\ x \equiv -1 \mod q \end{cases} \quad \begin{cases} x \equiv 1 \mod p \\ x \equiv -1 \mod q \end{cases} \quad \begin{cases} x \equiv -1 \mod p \\ x \equiv 1 \mod q \end{cases}.$$

The first set of equations has the solution $x \equiv 1 \mod pq$, and the second set has the solution $x \equiv -1$. If we denote the solution of the third set by ϵ, then the solution of the fourth set is $-\epsilon$.

(ii) The argument of part (i) shows that the equation $x^2 \equiv a^2 \mod pq$ has at most four distinct roots. By inspection a, $-a$, ϵa and $-\epsilon a$ are distinct roots, so we are done.   ∎

Exercise 5.5.2 *Suppose, as before, that p and q are distinct odd primes.*

(i) *If $a \not\equiv 0 \mod p$ and $a \equiv 0 \mod q$, show that the equation $x^2 \equiv a^2 \mod pq$ has two roots and identify them.*

(ii) *What happens if $a \equiv 0 \mod p$ and $a \equiv 0 \mod q$ and why?*

(iii) *Show that exactly $(p + 1)(q + 1)/4$ elements of $\mathbb{Z}_{pq}$ are squares.*

Exercise 5.5.3 *If $p = 7$ and $q = 13$, use Bézout's algorithm to compute the two solutions of $x^2 \equiv 1 \mod pq$ with $x \not\equiv \pm 1$.*
[If you dislike toy examples, just choose larger primes.]

We now leave the Greek world, where numbers are studied for their own sake, and re-enter the Egyptian world, where numbers have a practical purpose. Just like the Egyptian scribes, we will worry about the difficulty or ease of computations but, for us, a calculation will be considered easy if a computer can do it in a short time (that is to say, in a few steps) and difficult if it requires a long time (that is to say, it requires many steps). Actually, we will go a little further and consider that a calculation is easy if, when we choose the input according to some specified random method, the probability our computer will fail to complete the required computation in a short time is very small indeed.[9]

We now consider large primes p and q (think of numbers of the size of 10^{400}) such that $p = 4n + 3$ and $q = 4m + 3$ with m and n non-negative integers. We claim that the following two problems are essentially of equivalent difficulty. Here, we know the value of $N = pq$, but are not told p or q.

Problem A *(Factorisation) Given N, find p and q.*

Problem B *(Square roots) Given a, which we know to be a square modulo N, find x such that $x^2 \equiv a \mod N$.*

─────────────────────

[9] Say, less than the chance of you being struck by a meteorite in the next twenty-four hours.

Lemma 5.5.4 *If we can solve Problem A, we can solve problem B easily.*

Proof. Solving Problem A gives us p and q. By Lemma 5.2.13 (iii), $x \equiv \pm a^{(p+1)/4} \mod p$ and $x \equiv \pm a^{(q+1)/4} \mod q$. The algorithm for the Chinese remainder theorem now gives all the square roots. ∎

Lemma 5.5.5 *If we can solve Problem B, then there is an easy procedure which solves Problem A with very high probability in a small number of attempts.*

Proof. Suppose that we have a device[10] which solves Problem B, that is to say, if we feed the device a number a, which we know to be a square modulo $N = pq$, it will deliver an x such that $x^2 \equiv a \mod N$.

We choose an integer a at random with $1 \le a \le N - 1$ (all choices being equally likely). By applying Euclid's algorithm to N and a, either we learn that a is not coprime to N (incredibly unlikely if p and q are really large, but in this case Euclid's algorithm gives p or q as the highest common factor of a and N) or we obtain (via Bézout's algorithm) c such that $ca \equiv 1 \mod N$.

We now feed the machine $b = a^2$. The machine will return an x such that $x^2 \equiv b \mod N$, but, since a was chosen at random, x is equally likely to be a, $-a$, ϵa or $-\epsilon a$, where ϵ is defined as in Lemma 5.5.1. If $x \equiv a$ or $x \equiv -a$ mod N, we have not learned anything, but, with probability $1/2$, $x \not\equiv \pm a$ mod N. In this second case, we calculate $\eta \equiv cx$ and observe that $\eta \equiv \epsilon$ or $\eta \equiv -\epsilon \mod N$.

By interchanging the names of p and q, if necessary, we may suppose that $\eta \equiv \epsilon \mod N$. Then

$$\eta \equiv 1 \mod p$$
$$\eta \equiv -1 \mod q.$$

Since q divides $\eta + 1$, but p does not, the highest common factor of $\eta + 1$ and N is q. Euclid's algorithm applied to N and $\eta + 1$ now gives q. Knowing N and q, we obtain p.

If our attempt has failed because the machine returned a known root, we repeat the attempt with another randomly chosen a. The probability that n successive attempts will fail is 2^{-n}, which becomes very small very rapidly as n increases. ∎

Exercise 5.5.6 *Suppose $N = 437$ and you know that N is the product of two primes of the form $4n + 3$. Starting from the observation that $112^2 \equiv 308 \equiv$*

---

[10] In standard mathematical terminology, an *oracle.*

302^2 mod 437, *carry out the calculations suggested to find the prime factors of 437.*

Exercise 5.5.7 *Check, if you do not already know this useful fact, that* $2^{-10} <$ 10^{-3} *(and the associated remark that* $2^{10} \approx 10^3$ *). Conclude that the probability the scheme outlined in Lemma 5.5.5 will fail to factorise* N *in 400 attempts is less than* 10^{-120}.

Exercise 5.5.8 *The algorithms described above all rely on the Euclidean algorithm, so we need to check that the algorithm is fast. Recall that starting with a pair* (u, v) *with* u *and* v *integers and* $u \geq v \geq 1$, *we either stop or produce a pair* (u', v') *with* $u' = v$

$$u = ku' + v',$$

where $k \geq 1$. *If we now make the next step, we either stop or produce a pair* (u'', v'') *with* $u'' = v'$ *and*

$$u' = k'u'' + v'',$$

where $k' \geq 1$.

   Show that $u/2 > u''$. *Deduce that, if we start Euclid's algorithm with a pair* (U, V) *where* $U \approx 10^{3m}$, *then the process will terminate in less than about* $20m$ *steps. The 'pulveriser' deserves its nickname.*

   Why does the fact that Euclid's algorithm is fast tell us that Bézout's algorithm is fast?

Exercise 5.5.9 *We have seen how to find the square roots of a modulo* p, *where* p *is a prime and* $p = 4n+3$, *using Lemma 5.2.13 (iii) and computing* $a^{(p+1)/2}$ *modulo* p. *On the face of it, this involves an immense amount of calculation, but we can use a simple and very old trick to get around this. Suppose* $r <$ 2^{m+1}. *Then we can write* r *in binary as*

$$r = \zeta_0 + \zeta_1 2 + \zeta_2 2^2 + \ldots \zeta_m 2^m,$$

with $\zeta_j \in \{0, 1\}$ *for* $0 \leq j \leq m$. *Successive squaring allows us to compute* a^2, $a^4 = (a^2)^2, \ldots, a^{2^k} = (a^{2^{k-1}})^2, \ldots, a^{2^m} = (a^{2^{m-1}})^2$. *Since*

$$a^r = a^{\zeta_0 2^0} \times a^{\zeta_1 2^1} \ldots a^{\zeta_k 2^k} \times \ldots a^{\zeta_m 2^m}$$

we obtain a^r *by multiplying together those* a^{2^k} *with* $\zeta_k = 1$. *In standard notation,*

$$a^r = \prod_{\zeta_k=1} a^{2^k}.$$

(i) *Check that this procedure only requires at most* 2m *multiplications.*
(ii) *Compute* 7^{100} *modulo* 23.

The equivalence of problems A and B forms the basis for a very good secret code. The Rabin code works as follows.

Step 1 Choose two very large prime numbers p and q with $p \equiv 3$ and $q \equiv 3 \mod 4$ according to some random means. (There are non-trivial theorems which show that this is quite easy to do.)

Step 2 Keep the values of p and q secret, but announce the value of $N = pq$ to the world.

Step 3 Anyone with a message for you converts it, according to some standard method, into integers $n_1, n_2, \ldots n_k$ with $0 \leq n_j < N$ for $1 \leq j \leq N$. They then send you $m_1, m_2, \ldots m_k$ with $m_j \equiv n_j^2 \mod N$.

Step 4 Since you know p and q, you can find the four possible solutions of $x_j \equiv m_j^2 \mod N$. Provided that simple precautions are taken,[11] the probability that more than one of these roots could be the correct one will be vanishingly small. Thus you can recover the original message.

Now suppose that someone else has got hold of the coded messages m_j. If she can decode them on a regular basis, then she can extract square roots modulo N. But, if she can extract square roots modulo N, then, as we have seen, she can factorise N. If we change our primes frequently, then anyone who can decode our messages without knowing our secret primes is in possession of a method of factorising products of two large primes. Since mathematicians have been seeking such a method for the last 300 years (and with particular fervour for the last 40 years), our prospective code-breaker must be very clever indeed.

If you use a pre-1970 code, then anyone who has sufficient information to translate messages into that code will also have sufficient information to decode coded messages. The Rabin and related codes have the property, very valuable in the age of internet banking, that everybody can translate messages into code, but only those in possession of a particular secret (here the factorisation of N) can decode a coded message.

For various reasons, the most used code of this type is not the Rabin code, but another code called the RSA[12] code. The security of the RSA code also depends on the presumed difficulty of the factorisation problem.

Of course, new mathematical ideas (or new technology, such as quantum computation) may mean that the factorisation problem ceases to be hard. This

---

[11] A particularly clever way of avoiding ambiguity was suggests by Williams. When this idea is used, people sometimes talk about the Rabin–Williams code.

[12] Rivest, Shamir, Adleman.

means that 'factorisation codes' should only be used for secret messages which (like most secret messages[13]) need remain secret only for a fairly short time. A substantial level of security is provided by the high probability that any mathematician who solves the factorisation problem will prefer fame and a professorship at her favourite university to prolonged and difficult negotiations with rather unsavoury characters.

The reader who is tempted to use this as an example of an unexpected benefit of abstract thought should keep two things in mind.

The first is that, by attempting to justify pure mathematics through its usefulness, you play straight into the hands of crude utilitarians. You should remember Hardy's sour remark[14], first made in 1915 and repeated by him in 1940, that 'a science is said to be useful if its development tends to accentuate the existing inequalities of wealth, or more directly promotes the destruction of human life.'

At a second and lower level, you should imagine the effect on a grant-giving body of the argument that some piece of research might be useful in a couple of hundred years' time.

Exercise 5.5.10 *This exercise simply shows that there are an infinity of primes of the form $4n + 3$, and not, as stated (correctly) in Step 1 of our description of the Rabin code, that they are easy to find.*

(i) Suppose that $q_j \equiv 1 \mod 4$ for $1 \le j \le r$. Show that $q_1 q_2 \ldots q_r \equiv 1 \mod 4$. Deduce that, if M is a non-negative integer, then $4M + 3$ must have a prime factor p with $p \equiv 3 \mod 4$.

(ii) Use the argument of Theorem 4.4.9 to show that there are an infinity of primes of the form $4n + 3$.

Exercise 5.5.11 *Use Lemma 5.2.12 to show that $4M^2 + 1$ is not divisible by any prime of the form $4n + 3$. By considering integers of the form*

$$4(p_1 p_2 \ldots p_m)^2 + 1$$

and using the argument of Theorem 4.4.9, show that there are an infinity of primes of the form $4n + 1$.

Exercise 5.5.12 *I announce that I shall be using the Rabin scheme with modulus N. My agent in X'Dofdro sends me a message m (with $1 \le m \le N - 1$) encoded in the requisite form. Unfortunately, my white Persian cat eats the piece of paper on which the prime factors of N are recorded, so I am unable*

---

[13] 'Do not accept any offer under two million.'; 'The fleet will sail at dawn.'

[14] 'A conscious rhetorical flourish, but perhaps excusable at the time when it was written' [13].

to decipher it. I therefore find a new pair of primes and announce that I shall be using the Rabin scheme with modulus $N' > N$. My agent now recodes the message and sends it to me again.

The dreaded SNDO of X'Dofdro intercept both code messages. Show that they can find m. Can they decipher any other messages sent to me using only one of the coding schemes?

6

Axioms for the Natural Numbers

6.1 The Peano Axioms

We have seen several systems that resemble the natural numbers in one way or another and we shall see several more. What is special about the natural numbers? It cannot be the algebraic rules that we started with, since these are shared, to a greater or lesser extent, with the other systems. However, the principle of induction does appear to be special and it is the wish to recover the principle of induction which lies behind our choice of axioms for the natural numbers.

We shall use the Peano axioms.[1] They describe a set $\mathbb{N}^+$ which we call the strictly positive integers, a function $S : \mathbb{N}^+ \to \mathbb{N}^+$ (we call $S(x)$ the *successor* of x) and a *base number* $1 \in \mathbb{N}^+$ satisfying the following three conditions:

(P1) *If* $x \in \mathbb{N}^+$, *then* $S(x) \neq 1$.

    (The base number 1 is not the successor of any natural number.)

(P2) *If* x, $y \in \mathbb{N}^+$ *and* $S(x) = S(y)$, *then* $x = y$.

    (Two natural numbers with the same successor are equal.)

(P3) *If* E *is a set of natural numbers such that* $1 \in E$ *and, whenever* $x \in E$, *we have* $S(x) \in E$, *then* $E = \mathbb{N}^+$.

    (An axiom of induction. If a collection of natural numbers contains 1 and with every element its successor, then it contains all the natural numbers.)

The reader should pause and think whether she agrees that these are properties we would wish the natural numbers to have.

---

[1] Sometimes called the Peano–Dedekind axioms, since they are a modification of axioms found by Dedekind.

Exercise 6.1.1 *It is often useful when trying to understand the meaning of a set of axioms to consider systems which obey some of the axioms, but not others. Check the following statements.*

(i) *Let $\mathbb{N}_1^+$ consist of the single element 1 and let $S_1(1) = 1$. The system satisfies all the axioms except (P1).*

(ii) *Let $\mathbb{N}_2^+$ consist only of the elements 1 and 2 and let $S_2(1) = S_2(2) = 2$. The system satisfies all the axioms except (P2).*

(iii) *Let $\mathbb{N}_3^+$ consist of all strictly positive integers together with the rational numbers of the form $n + 1/2$ with n an integer and let $S_3(a) = a + 1$. The system satisfies all the axioms except (P3).*

Peano's axioms seem to capture the properties of natural numbers as 'ordered numbers'. Are they powerful enough to give the standard algebraic operations and relations? We shall see that the answer is yes, but, as might be expected, we shall need to proceed very carefully to show this. If the reader thinks of a very heavy freight train very slowly gathering speed from a standing start, she will have a good image of the succession of results that we need to prove on the way to Theorem 6.2.8. Once we get to that point, we can relax.

Throughout, we only assume Peano's axioms. We write $x' = S(x)$.

We wish to define addition, that is to say, we wish to define $x + y$ for natural numbers x and y. Unfortunately, the expression $x + y$ involves 'two variables x and y' and our axioms only refer to 'one variable x'. We get round the problem by first 'fixing x and allowing y to vary' (see the uniqueness part of the proof that follows) and then 'fixing y and allowing x to vary' (see the existence part of the proof).

Theorem 6.1.2 *Let $x \in \mathbb{N}^+$. There is a unique function $\theta_x : \mathbb{N}^+ \to \mathbb{N}^+$ satisfying the following conditions.*

(a) $\theta_x(1) = x'$.

(b) $\theta_x(y') = \big(\theta_x(y)\big)'$ *for all natural numbers y.*

Proof. Uniqueness We first prove that there is at most one function with the desired properties. Suppose that θ_x and ϕ_x have the properties stated, so that

(a) $\theta_x(1) = \phi_x(1) = x'$,

(b) $\theta_x(y') = \big(\theta_x(y)\big)'$ and $\phi_x(y') = \big(\phi_x(y)\big)'$ for all natural numbers y.

Let E be the set of natural numbers y such that $\phi_x(y) = \theta_x(y)$. Condition (a) tells us that

$$\phi_x(1) = x' = \theta_x(1),$$

so $1 \in E$. On the other hand, if $y \in E$, condition (b) tells us that

$$\phi_x(y') = \left(\phi_x(y)\right)' = \left(\theta_x(y)\right)' = \theta_x(y'),$$

so $y' \in E$. The induction axiom (P3) now tells us that $E = \mathbb{N}^+$, which is what we wished to prove.

Existence Let E be the collection of natural numbers x such that we can define θ_x with properties (a) and (b). Observe that, if we set $\tilde{\theta}_1(y) = y'$, then

(a) $\tilde{\theta}_1(1) = 1'$, and
(b) $\tilde{\theta}_1(y') = (y')' = \left(\tilde{\theta}_1(y)\right)'$ for all natural numbers y.

Thus $1 \in E$.

We now suppose $x \in E$ and so there exists a function θ_x with properties (a) and (b). Observe that, if we set $\tilde{\theta}_{x'}(y) = \left(\theta_x(y)\right)'$, then

(a') $\tilde{\theta}_{x'}(1) = \left(\theta_x(1)\right)' = (x')'$, and
(b') $\tilde{\theta}_{x'}(y') = \left(\theta_x(y')\right)' = \left(\left(\theta_x(y)\right)'\right)' = \left(\tilde{\theta}_{x'}(y)\right)'$ for all natural numbers y.

Thus $x' \in E$. The induction axiom (P3) now tells us that $E = \mathbb{N}^+$, which is what we wished to prove. ∎

Lemma 6.1.3 *The functions θ_x obtained in Theorem 6.1.2 satisfy the equation*

$$\theta_1(y) = y'$$

for all natural numbers y and the equation

$$\theta_{x'}(y) = \theta_x(y')$$

for all natural numbers x and y.

Proof. Let E be the collection of natural numbers y such that $\theta_1(y) = y'$. We have $\theta_1(1) = 1'$, so $1 \in E$. If $y \in E$, then, by the definition of θ_1,

$$\theta_1(y') = \left(\theta_1(y)\right)' = (y')',$$

so $y' \in E$. The induction axiom (P3) now tells us that $E = \mathbb{N}^+$ and so $\theta_1(y) = y'$ for all y.

Let x be a natural number and let E_x be the collection of natural numbers y such that

$$\theta_{x'}(y) = \theta_x(y').$$

We know that

$$\theta_{x'}(1) = x'' = \left(\theta_x(1)\right)' = \theta_x(1'),$$

so $1 \in E_x$. Further, if $y \in E_x$, then

$$\theta_{x'}(y') = \left(\theta_{x'}(y)\right)' = \left(\theta_x(y')\right)' = \theta_x(y''),$$

so $y' \in E_x$. Thus, by the induction axiom (P3), $E_x = \mathbb{N}^+$ and this is equivalent to the statement of the second claim of our theorem. ∎

We can now remove the asymmetry of x and y by setting $x + y = \theta_x(y)$. The definition given in Theorem 6.1.2 and the conclusion of Lemma 6.1.3 show that

$$x + 1 = \theta_x(1) = x' = \theta_1(x) = 1 + x$$

and

$$x + y' = \theta_x(y') = (x + y)' = \theta_{x'}(y) = x' + y,$$

or, more succinctly,

$$x + 1 = x' = 1 + x \text{ and } x + y' = (x + y)' = x' + y \qquad \bigstar$$

for all natural numbers x and y.

Initially, proofs like the one just given of Theorem 6.1.2 look totally opaque. After some practice, the beginner may swing to the opposite opinion. 'Surely this proof is trivial? You just define the appropriate E and perform the appropriate verifications.' It is worth pointing out that the verifications have to be carried out in the right order. More importantly, much of the skill of Peano and his predecessors,[2] lay in choosing a collection of axioms and definitions which make the proofs run easily.

Once we have got this far, the commutative and associative laws of addition are easy to obtain.

Lemma 6.1.4 *If a, b and c are natural numbers, the following equalities hold.*

(i) $a + b = b + a$. *(Commutative law)*
(ii) $(a + b) + c = a + (b + c)$. *(Associative law)*

Proof. (i) Fix a and let E be the set of natural numbers b with $a + b = b + a$. By $\bigstar$,

$$a + 1 = 1 + a,$$

so $1 \in E$. If $b \in E$, then, using $\bigstar$ again,

$$a + b' = (a + b)' = (b + a)' = b' + a,$$

---

[2] Including Grassmann, whose work on this, as well as his even more important work on what we now call vectors (see page 230), was considerably ahead of its time.

so $b' \in E$. The induction axiom (P3) now tells us that $E = \mathbb{N}^+$ and the required result follows.

(ii) Fix a and b and let E be the set of natural numbers c with $(a + b) + c = a + (b + c)$. By ★,

$$(a + b) + 1 = (a + b)' = a + b' = a + (b + 1),$$

so $1 \in E$. If $c \in E$, then, using ★ again,

$$(a+b)+c' = \big((a+b)+c\big)' = \big(a+(b+c)\big)' = a+(b+c)' = a+(b+c'),$$

so $c' \in E$. The induction axiom (P3) now tells us that $E = \mathbb{N}^+$ and the required result follows.   ■

As we shall see in Section 6.3, we can obtain multiplication in a similar manner. For the moment, we turn to the more challenging task of obtaining the usual order $<$ on $\mathbb{N}^+$.

6.2 Order

It is clear that the notion of successor in the Peano axioms is related to the standard notion of order $x < y$, but, if the reader reflects, she will see that it is not that easy to derive the one from the other. We looked at addition first to provide an appropriate tool. The reader may also have noticed that, in deriving addition, we made use only of the axiom (P3) (see Exercise 6.3.10). Our treatment of order will require all the axioms.

We make a start with some fairly easy observations.

Lemma 6.2.1

(i) *If $x \neq y$, then $x' \neq y'$. (Unequal natural numbers have unequal successors.)*

(ii) *$x \neq x'$. (No natural number equals its successor.)*

(iii) *If $x \neq 1$, then there exists a unique y with $x = y'$. (Every natural number, except 1, has a unique predecessor.)*

Proof. (i) If $x' = y'$, then, by (P2), $x = y$.

(ii) Let E be the collection of natural numbers x such that $x \neq x'$. By (P1), $1 \neq 1'$, so $1 \in E$. If $x \in E$, then $x \neq x'$ and, by (i), $x' \neq (x')'$, so $x' \in E$. The induction axiom (P3) now tells us that $E = \mathbb{N}^+$, which is what we wished to prove.

(iii) Uniqueness is guaranteed by (P2), so we need only prove existence. Let E be the collection of natural numbers consisting of 1 together with all those x such that there exists a y with $y' = x$. We have $1 \in E$ by definition. Now suppose $x \in E$. Automatically, x' is the successor of x, so $x' \in E$. The induction axiom (P3) now tells us that $E = \mathbb{N}^+$, which is what we wished to prove. ∎

Results like those of Lemma 6.2.1 are easy to prove, but, if you forget to prove them, there are problems later. We now combine the axioms with the results on addition (see, particularly, the formulae labelled ★ on page 112). The next two lemmas say, in effect, that we cannot go round in circles. (Contrast modular arithmetic.)

Lemma 6.2.2 *If x and y are natural numbers, then*

$$y \neq x + y.$$

Proof. Fix x and let E be the collection of natural numbers y such that $y \neq x + y$. We know, by axiom (P1), that 1 is not a successor, so

$$1 \neq x' = x + 1.$$

Thus $1 \in E$.

On the other hand, if $y \in E$, then

$$y \neq x + y.$$

Lemma 6.2.1 (i) now tells us that

$$y' \neq (x + y)' = x + y',$$

so $y' \in E$. The induction axiom (P3) now tells us that $E = \mathbb{N}^+$, which is what we wished to prove. ∎

Lemma 6.2.3 *If x, y and z are natural numbers with $x + y = x + z$, then $y = z$.*

Proof. We prove an equivalent statement: If a, b and c are natural numbers with $b \neq c$, then

$$a + b \neq a + c.$$

To this end, fix b and c with $b \neq c$ and let E be the collection of natural numbers a such that $a + b \neq a + c$. By ★ and Lemma 6.2.1 (i), we know that

$$1 + b = b' \neq c' = 1 + c,$$

so $1 \in E$.

On the other hand, if $a \in E$, we have, using ★ and Lemma 6.2.1 (i) again,

$$a' + b = (a + b)' \neq (a + c)' = a' + c,$$

so $a' \in E$. The induction axiom (P3) now tells us that $E = \mathbb{N}^+$, which is what we wished to prove. ∎

We can now prove a key result.

Theorem 6.2.4 *Suppose x and y are natural numbers. Then exactly one of the following statements must be true.*

(1) $x = y$.
(2) There is a natural number u such that $x + u = y$.
(3) There is a natural number v such that $y + v = x$.

Proof. We begin by proving that at most one of the three statements must be true. Lemma 6.2.2 tells us that (1) and (2) cannot both be true and that (1) and (3) cannot both be true. Now suppose (2) and (3) are both true. We then have

$$x + (u + v) = (x + u) + v = y + v = x,$$

which contradicts Lemma 6.2.2 again.

We now prove that one of the statements must be true. As usual, we fix x and consider the collection E_x of natural numbers y for which at least one (and so, by the first paragraph, exactly one) of the statements are true.

We note first that, if $x = 1$, then, looking at statement (1), we see that $1 \in E_1$. If $x \neq 1$, then Lemma 6.2.1 (iii) tells us that $x = u'$ for some $u \in \mathbb{N}^+$, so $x = u + 1 = 1 + u$ and, looking at statement (3), we see that $1 \in E_x$.

Now suppose that $y \in E_x$.
Case 1 If $x = y$, then $y' = y + 1 = x + 1$ and statement (2) holds with y replaced by y' and $u = 1$.
Case 2 If $x + u = y$, for some u, then

$$x + u' = (x + u)' = y',$$

so statement (2) holds with y replaced by y' and u by u'.
Case 3 If $x = y + v$ for some v, then there are two possibilities. The first is that $v = 1$ so that

$$x = y + 1 = y'$$

and statement (1) holds with y replaced by y'. The second is that $v \neq 1$ and so, again by Lemma 6.2.1 (iii), there is a w such that $v = w'$. In this case

$$x = y + v = y + w' = (y + w)' = y' + w,$$

so statement (3) holds with y replaced by y' and v by w.

In all three cases, $y' \in E_x$. The induction axiom (P3) now tells us that we have $E_x = \mathbb{N}^+$ as required.   ∎

The way is now clear to define the notion of inequality exactly as we did on page 18.

Definition 6.2.5 *If a and b are natural numbers, we write $a > b$ if and only if there exists a natural number c with $b + c = a$.*

We now obtain trichotomy not as a *rule* but as a *theorem*.

Theorem 6.2.6 [Trichotomy] *If a and b are natural numbers, then exactly one of the following conditions holds: $a > b$ or $b > a$ or $a = b$.*

Proof. This is the content of Theorem 6.2.4.   ∎

We have now recovered all our rules for addition and inequality with the exception of our golden key, the least member principle. However, it is pretty clear that a careful reuse of the argument we used to prove Theorem 4.1.7 will give us the required rule.

We need a preliminary lemma.

Lemma 6.2.7

(i) *If x is a natural number, then $x \geq 1$.*
(ii) *If $y > x$, then $y \geq x + 1$.*
(iii) *If $a' > b$, then $a \geq b$.*

Proof. (i) Either $x = 1$ or, by Lemma 6.2.1 (iii), $x = y'$ for some y. Then

$$x = y + 1 = 1 + y$$

and $x > 1$.

(ii) By definition, $y = x + u$ for some u. By part (i), $u \geq 1$, so either $u = 1$ or $u > 1$. If $u = 1$, then $y = x + 1$ and we are done. If $u > 1$, then $u = v + 1$ for some v and

$$y = x + (v + 1) = x + (1 + v) = (x + 1) + v,$$

so $y > x + 1$.

(iii) Suppose that $b > a$. Then, by part (ii), $b \geq a + 1 = a'$, which trichotomy tells us is impossible.   ∎

Theorem 6.2.8 [Least member principle] *If A is a non-empty collection of natural numbers, we can find an $a_0 \in A$ such that $a \geq a_0$ for all $a \in A$.*

Proof. We call the a_0, described in the statement of our theorem, a least member of A. Let E be the set of natural numbers m such that, if $m \geq r$ and $r \in A$ (a collection of natural numbers), then A has a least member.

If $m = 1$ and $m \geq r$, then $1 \geq r$ and, by Lemma 6.2.7 (i), $r \geq 1$. By trichotomy $r = 1$. If $1 \in A$, then, by Lemma 6.2.7 (i) again, 1 is a least member of A. Thus $1 \in E$.

Now suppose $m \in E$, $m' \geq r$ and $r \in A$. There are two possibilities: either there does not exist an $s \in A$ with $m' > s$ or there does.

If there does not exist an $s \in A$ with $m' > s$, then $m' = r$ and m' is itself a least member of A.

If there does exist an $s \in A$ with $m' > s$, then by Lemma 6.2.7 (iii), $m \geq s$. Since $m \in E$, it follows that A has a least member.

In either case, A has a least member. We have thus shown that $m' \in E$. The induction axiom (P3) now tells us that $E = \mathbb{N}^+$. Since any non-empty A contains some r we can set $r = m$ to obtain the required result. ∎

The easy argument of Lemma 4.1.1 shows that there is exactly one least member, so we may refer to *the* least member.

6.3 Conclusion of the Argument

We have done the hard part of the argument, but there remain various tasks to perform. The introduction of multiplication follows the same pattern as that of addition.

Theorem 6.3.1

(i) Let $x \in \mathbb{N}^+$. There is a unique function $\mu_x : \mathbb{N}^+ \to \mathbb{N}^+$ satisfying the following conditions.

(a) $\mu_x(1) = x$.

(b) $\mu_x(y') = \mu_x(y) + x$ for all natural numbers y.

(ii) Further, this function satisfies the equation

$$\mu_1(y) = y$$

for all natural numbers y and the equation

$$\mu_{x'}(y) = \mu_x(y) + y$$

for all natural numbers x and y.

Proof. This follows the proof of Theorem 6.1.2 and Lemma 6.1.3 and is left to the reader.   ∎

Exercise 6.3.2 *Write out the proof of part (i) of Theorem 6.3.1.*
[*This is quite easy to do by copying the proof of Theorem 6.1.2 and making the appropriate changes. However, it will be more profitable if, from time to time, the reader thinks about what is happening.*]

Exercise 6.3.3 *Write out the proof of part (ii) of Theorem 6.3.1.*

We can now remove the asymmetry of x and y by setting $x \times y = \mu_x(y)$ and observing that

$$x \times 1 = x = 1 \times x, \ x \times y' = (x \times y) + x, \ x' \times y = (x \times y) + y \quad ★$$

for all natural numbers x and y. Notice that ★ contains the 'one times multiplication table'.

Lemma 6.3.4 *If a is a natural number, then $1 \times a = a$.*

We can now prove the remaining rules discussed in Section 1.1.

Lemma 6.3.5 *Let a, b and c be natural numbers. Then the following results hold.*

(i) $a \times (b + c) = (a \times b) + (a \times c)$. (*Distributive law*)
(ii) $a \times b = b \times a$. (*Commutative law for multiplication*)
(iii) $(a \times b) \times c = a \times (b \times c)$. (*Associative law for multiplication*)

Proof. (i) Fix a and b. Let E be the collection of natural numbers c such that

$$a \times (b + c) = (a \times b) + (a \times c).$$

We note that

$$a \times (b + 1) = a \times b' = (a \times b) + a = (a \times b) + (a \times 1),$$

so $1 \in E$.

On the other hand, if $c \in E$, then, using the associative property of addition,

$$\begin{aligned}
a \times (b + c') &= a \times (b + c)' = \big(a \times (b + c)\big) + a \\
&= \big((a \times b) + (a \times c)\big) + a = (a \times b) + \big((a \times c) + a\big) \\
&= (a \times b) + (a \times c'),
\end{aligned}$$

so $c' \in E$. The induction axiom (P3) now tells us that $E = \mathbb{N}^+$, which is what we wished to prove.

(ii) [Observe that we shall use (i) both here and in the proof of (iii).] Fix a and let E be the set of natural numbers b with $a \times b = b \times a$. By ★,

$$a \times 1 = 1 \times a$$

so $1 \in E$. If $b \in E$, then, using (i), ★, the commutative property of addition and the relation $x' = x + 1$,

$$a \times b' = a \times (b + 1) = (a \times b) + (a \times 1) = (a \times b) + a$$
$$= (b \times a) + a = a + (b \times a) = (1 \times a) + (b \times a)$$
$$= (1 + b) \times a = (b + 1) \times a = b' \times a$$

so $b' \in E$. The induction axiom (P3) now tells us that $E = \mathbb{N}^+$ and the required result follows.

(iii) Fix a and b. Let E be the collection of natural numbers c such that

$$(a \times b) \times c = a \times (b \times c).$$

We note that

$$(a \times b) \times 1 = a \times b = a \times (b \times 1),$$

so $1 \in E$.

On the other hand, if $c \in E$, then, using (i),

$$a \times (b \times c') = a \times ((b \times c) + b) = (a \times (b \times c)) + (a \times b)$$
$$= ((a \times b) \times c) + (a \times b) = (a \times b) \times c',$$

so $c' \in E$. The induction axiom (P3) now tells us that $E = \mathbb{N}^+$, which is what we wished to prove. ∎

Exercise 6.3.6 *Cast a quick eye over the mathematical part of Section 1.3 to check that the results of this chapter match up with what is required there.*

We complete the discussion by showing that the Peano system is unique (subject to the standard caveat *up to isomorphism*). The reader is reminded that (as our discussion of modular arithmetic shows) the axioms for a field (see Definition 5.1.1) do not define a unique system. Later we shall see that, even if we add order axioms, there are many ordered fields.

Theorem 6.3.7 *Suppose that A is a set, S a successor function on A and α a base number such that the Peano axioms are obeyed and that (B, T, β) is another such system obeying the Peano axioms. Then there is a bijective function $f : A \to B$ such that $f(\alpha) = \beta$ and $T(f(a)) = f(S(a))$ for all $a \in A$.*

The reader should pause to think whether, in her opinion, the theorem just stated shows that (so far as successor properties go) the Peano system is essentially unique.

Before beginning the proof of Theorem 6.3.7 we observe that, since any system obeying the Peano axioms obeys least element principle, we can obtain all the results of Chapter 4 and, in particular, Theorem 4.2.1 which we restate in a more abstract form.

Theorem 6.3.8 *Suppose that A is a set, S a successor function on A and α a base number such that the Peano axioms are obeyed. Let X be a set with $x_\alpha \in X$. Suppose that whenever $a \in A$ we have a function $g_a : X \to X$. Then there exists a unique $f : A \to X$ such that $f(\alpha) = x_\alpha$ and $f(S(a)) = g_a(f(a))$ for all $a \in A$.*

Exercise 6.3.9 *Cast a quick eye over the first two sections of Chapter 4 and check that Theorem 6.3.8 is indeed a restatement of Theorem 4.2.1.*

Proof of Theorem 6.3.7. Since (A, S, α) obeys the Peano axioms, Theorem 6.3.8 allows us to define a function $f : A \to B$ inductively by the formulae

$$f(\alpha) = \beta \text{ and } f(S(a)) = T(f(a)).$$

Similarly, we can define $g : B \to A$ inductively by the formulae

$$g(\beta) = \alpha \text{ and } g(T(b)) = S(g(b)).$$

I claim that $g(f(a)) = a$ for all $a \in A$. To show this, let us write E for the collection of $a \in A$ such that $g(f(a)) = a$. We observe that

$$g(f(\alpha)) = g(\beta) = \alpha$$

and that, if $a \in E$, then

$$g(f(S(a))) = g(T(f(a))) = S(g(f(a))) = S(a),$$

so $S(a) \in E$. The induction axiom (P3) for (A, T, α) now tells us that $E = A$, which is what we wished to prove.

The same argument tells us that $f(g(b)) = b$ for all $b \in B$. Thus [see Exercise 3.4.11 (ii)] f is a bijection. ∎

Exercise 6.3.10 *I said earlier that our definition of addition only made use of axiom (P3). Now that we have established the properties of $\mathbb{N}^+$, we may use results from Parts I and II of this book to illustrate the point.*

Figure 6.1. *Can't sleep*, an xkcd cartoon (https://xkcd.com).

Let q be an integer with $q \geq 2$. We consider the set $\mathbb{Z}_q$ of equivalence classes $[r]$ modulo q (see Figure 6.1 and the discussion starting with Exercise 5.1.10). If we define a function $S : \mathbb{Z}_q \to \mathbb{Z}_q$ by $S([x]) = [x + 1]$, note that $[1] \in \mathbb{Z}_q$ and show that the system obeys the following conditions.

(Q2) If $[x]$, $[y] \in \mathbb{Z}_q$ and $S([x]) = S([y])$, then $[x] = [y]$.

(Q3) If E is a subset of $\mathbb{Z}_q$ such that $[1] \in E$ and whenever $[x] \in E$, we have $S([x]) \in E$, then $E = \mathbb{Z}_q$.

Since we obtained addition for the Peano system only using (P3), it follows that we can use (Q3) to obtain the following analogue of Theorem 6.1.2. Let $[x] \in \mathbb{Z}_q$. There is a unique function $\phi_{[x]} : \mathbb{Z}_q \to \mathbb{Z}_q$ satisfying the following conditions.

(a) $\phi_{[x]}([1]) = S([x])$.

(b) $\phi_{[x]}\big(S([y])\big) = S\big(\phi_{[x]}([y])\big)$ for all $[y] \in \mathbb{Z}_q$.

Show that $\phi_{[x]}([y]) = [x] + [y]$, where '$+$' is our standard addition for modular arithmetic.

Obtain similar results for multiplication.

6.4 Order Numbers Can Be Used as Counting Numbers

If I look at a field, I can immediately spot if there are no sheep, one sheep or two sheep in it. If I look at a field with many sheep, I may be able to *estimate* the number of sheep in it, but the only way I know of finding the actual number of sheep in the field is to count them. How do we count sheep?

Typically, we count the sheep as they enter the field, counting 'First sheep in', 'Second sheep in', 'Third sheep in' and so on until we get to the last and nth sheep to enter. The sheep now mill around until we open the gate to let them out, counting 'First sheep out', 'Second sheep out', 'Third sheep out' and so on until we get to the last and mth sheep to leave. Does $m = n$? In the real world this is a matter of experiment.[3] In the world of mathematics it is a matter of proof.

The Peano axioms give us the kind of *order numbers* 'first element' 1, 'second element' $1' = 1 + 1$, 'third element' $1'' = (1 + 1) + 1$ and so on which are used when counting sheep. Our job is to show that we can use the Peano axioms to obtain a satisfactory way of counting collections of mathematical objects (that is to say, sets). In particular, we need to show that, however a set is presented to us, our counting system will always produce the same number. This is the content of Lemma 6.4.3, but we shall need to work quite hard to arrive there. Let us write F_n for the collection of natural numbers r with $r \leq n$. We write $2 = 1 + 1$.

Lemma 6.4.1

(i) *If there exist natural numbers m and n with $n > m$ and an injective function $f : F_n \to F_m$, then there exists an injective function $g : F_{m+1} \to F_m$.*

(ii) *If there exists a natural number m and an injective function $f : F_{m+2} \to F_{m+1}$, then there exists an injective function $g : F_{m+1} \to F_m$.*

(iii) *If n and m are natural numbers with $n > m$, then there does not exist an injective function $f : F_n \to F_m$.*

(iv) *If m and n are natural numbers, then there exists a bijective function $f : F_m \to F_n$ if and only if $m = n$.*

Proof. (i) We know (see Lemma 6.2.7 (ii)) that $n \geq m + 1$, so, if $r \in F_{m+1}$, it follows that $r \in F_n$. Set $g(r) = f(r)$ for all $r \in F_{m+1}$.

(ii) There are two possibilities. Either $f(r) \leq m$ for all $r \leq m + 1$ or there exists a $u \leq m + 1$ such that $f(u) = m + 1$.

If $f(r) \leq m$ for all $r \leq m + 1$, then we set $g(r) = f(r)$ for all $r \leq m + 1$ and g is automatically injective.

If there exists a $u \leq m+1$ such that $f(u) = m+1$, then, since f is injective, we know that $f(m + 2) = v$ with $v \neq m + 1$. By injectivity, $f(r) \neq v$ for all $r \in F_{m+2}$ with $r \neq m + 2$, that is to say, for all $r \in F_{m+1}$. If we now set

$$g(u) = v,$$
$$g(r) = f(r) \text{ for } r \neq u,$$

we obtain a well-defined injective function $g : F_{m+1} \to F_m$.

---

[3] 'They walked on, thinking of This and That, and by-and-by they came to an enchanted place on the very top of the Forest called Galleons Lap, which is sixty-something trees in a circle; and Christopher Robin knew that it was enchanted because nobody had ever been able to count whether it was sixty-three or sixty-four, not even when he tied a piece of string round each tree after he had counted it.' A. A. Milne, *The House at Pooh Corner*

(iii) By part (i), it is sufficient to prove the result when $n = m + 1$. To this end, let E be the collection of natural numbers such that there does not exist an injective function $f : F_{m+1} \to F_m$. We observe that, if f is a function from F_2 to F_1, then

$$f(1) = 1 = f(2),$$

so f is not injective. Thus $1 \in E$.

On the other hand, part (ii) tells us that, if $m \in E$, then $m + 1 \in E$. The axiom of induction now tells us that $E = \mathbb{N}^+$, which is what we wished to prove.

(iv) If $m \neq n$, then either $m > n$ or $n > m$. If $m > n$, we know that there is no injective and so no bijective function $f : F_n \to F_m$. If $n > m$, then, by the previous sentence, there is no bijective function $g : F_m \to F_n$ and so no bijective function $f : F_n \to F_m$.

If $n = m$, the identity function $f(r) = r$ gives a bijection between F_n and itself.   ∎

Exercise 6.4.2 *We obtained Lemma 6.4.1 (iv) by considering injective functions. Obtain the same result by considering surjective functions. (The most obvious way of doing this requires you to find 'surjective analogues' of the first three parts of the lemma.)*

We have an immediate corollary.

Lemma 6.4.3 *If A is a set and there exist bijections $f : F_n \to A$ and $g : F_m \to A$ for some natural numbers m and n, then $m = n$.*

Proof. Observe that the function $h : F_m \to F_n$ defined by $h(r) = f^{-1}\big(g(r)\big)$ is a bijection.   ∎

We can thus make the following definition.

Definition 6.4.4 *If A is a non-empty set such that there is a bijection $f : F_n \to A$ for some natural number n, we say that A is a finite set with n elements and write $|A| = n$. If no such bijection exists for any natural number, we say that A is infinite.*

We say that the empty set $\varnothing$ (that is to say, the set containing no elements) is finite and write $|\varnothing| = 0$.

If A is finite we call $|A|$ the number[4] of elements in A.

---

[4] Our definition of the natural numbers excludes 0, but, when talking about the empty set, it is useful to work with the non-negative integers $\mathbb{N}$. (See the footnote on page 50.)

Exercise 6.4.5

(i) If A and B are non-empty sets such that there exists a bijection $g : A \to B$, show that, if A is finite, then B is finite and $|A| = |B|$.

(ii) If A and B are non-empty finite sets and $|A| = |B|$ show that there exists a bijection $g : A \to B$.

Lemma 6.4.6 *The set $\mathbb{N}^+$ of natural numbers is infinite.*

Proof. Suppose, if possible, that $\mathbb{N}^+$ were finite. Then we could find a natural number n and a bijective function $f : F_n \to \mathbb{N}^+$. If we define $g : F_{n+1} \to F_n$, by $g(r) = f^{-1}(r)$ for $r \in F_{n+1}$, then g is injective, which we know to be impossible by Lemma 6.4.1 (iii). ∎

Remark The statement that $\mathbb{N}^+$ is infinite is unlikely to have surprised the reader, but nonetheless marks a significant intellectual parting of the ways. I can picture a field with 20 sheep, but, although I have frequently visited London and Paris, I find it very hard to picture 8 000 000 people. How then can I grasp the idea of an infinite collection of objects? It is perfectly reasonable to reject the existence, even in an abstract form, of collections of infinitely many objects and, for someone who rejects such collections, the Peano axiom system (along with the bulk of modern mathematics in its standard form) is thinly disguised nonsense. Like most mathematicians, I respect, but do not share, this point of view.

> If he's content with a vegetable love which would certainly not suit me
> Why what a most particularly pure young man, this pure young man must be.
>
> *W. S. Gilbert, Patience*

Exercise 6.4.7 *Check that the statement and proof of Theorem 4.4.9 are consistent with the definition of an infinite set just given.*

For the rest of this chapter, I shall assume a little more knowledge of set theory and its notation than elsewhere. We write $E \supseteq F$ to mean that F is a subset of E.

Lemma 6.4.8

(i) If n is a natural number and $F_n \supseteq B$, then B is finite and $n \geq |B|$.

(ii) If A is a finite set and $A \supseteq B$, then B is finite and $|A| \geq |B|$.

Proof. (i) The result is trivial when $B = \varnothing$, so we need only prove the result for B non-empty. Let E be the collection of natural numbers n such that, whenever B is a non-empty subset of F_n, B is finite with $n \geq |B|$.

We observe that, if $n = 1$, then the only non-empty subset of F_1 is F_1 itself, so $1 \in E$. Suppose now that $n \in E$ and $F_{n+1} \supseteq B$. If $n + 1 \notin B$, then B is a non-empty subset of F_n, so B is finite with $n + 1 \geq n \geq |B|$. If $n + 1 \in B$, we consider $B' = B \setminus \{n + 1\}$. If B' is empty, then $f : F_1 \to B$ defined by $f(1) = n + 1$ is a bijection, so $n + 1 \geq 1 = |B|$. If B' is non-empty, then, since B' is a subset of F_n, we know that there exists a natural number m with $n \geq m$ and a bijection $g : F_m \to B$. If we define $f(r) = g(r)$ for $m \geq r$ and $f(m + 1) = n + 1$, we obtain a bijection $f : F_{m+1} \to B$. Thus B is finite with $n + 1 \geq m + 1 = |B|$. The axiom of induction now tells us that $E = \mathbb{N}^+$, which is what we wished to prove.

(ii) The result is trivial if $B = \varnothing$, so we may suppose that B and so A are non-empty. Writing $n = |A|$, we know that there is a bijection $f : F_n \to A$. The set C consisting of those r with $f(r) \in B$ is a non-empty subset of F_n and so, by part (i), is finite with $n \geq |C|$. Since the function $g : C \to B$ given by $g(r) = f(r)$ for $r \in C$ is a bijection, we have B finite with $|B| = |C|$ and $|A| \geq |C| = |B|$. ∎

The reader who has got this far will almost certainly know that $A \cup B$ (the union of A and B) is the set consisting of all the x which belong to A or B (including those that belong to both) and $A \times B$ denotes the set of ordered pairs (a, b) with $a \in A$ and $b \in B$.

Lemma 6.4.9

(i) If A and B are disjoint finite sets (that is to say, A and B have no member in common), then $A \cup B$ is finite and $|A \cup B| = |A| + |B|$.

(ii) If A and B are finite sets, then $A \times B$ is finite and $|A \times B| = |A| \times |B|$.

Proof. (i) If A or B is the empty set, the result is trivial, so we may suppose $|A| = n$, $|B| = m$ for some natural numbers n and m. By definition, we can find bijective functions $f : A \to F_n$ and $g : B \to F_m$. If we set

$$h(r) = f(r) \qquad\qquad \text{if } n \geq r \geq 1,$$
$$h(n + u) = g(u) \qquad\qquad \text{if } m \geq u \geq 1,$$

then $h : F_{n+m} \to A \cup B$ is a bijective function and we are done.

(ii) If A or B is empty then, by definition, $A \times B$ is empty, so the result is trivial. Thus we may take A, B, n, m, f and g as in part (i). If we set

$$h(r) = \bigl(f(r), g(1)\bigr)$$

for $n \geq r \geq 1$, and

$$h(un + r) = \bigl(f(r), g(u + 1)\bigr)$$

for $n \geq r \geq 1$ and $m > u \geq 1$, then $h : F_{n \times m} \to A \times B$ is a bijective function and we are done.   ∎

Exercise 6.4.10 *Suppose that we omit the condition A and B disjoint in the statement of Lemma 6.4.9 (i). What can we say about $A \cup B$? Prove your answer.*

Exercise 6.4.11 *Use induction on m to show that, if n is a fixed natural number, there is a bijective function from F_{n^m} to the collection $\mathcal{F}_n(m)$ of functions $f : F_m \to F_n$.*

Deduce that, if A and B are non-empty finite sets and A^B is the set of all functions $g : B \to A$, then A^B is finite with $|A^B| = |A|^{|B|}$.

Exercise 6.4.12 *Suppose that A and B are finite non-empty sets with $|A| = |B|$. Use induction on the number of elements of A to prove the following two statements.*

(i) If $f : A \to B$ is injective, then f is bijective.
(ii) If $g : A \to B$ is surjective, then g is bijective.

Show, by means of examples, that the statements may be false if A and B are not finite.

Exercise 6.4.13 *Suppose that A is a non-empty finite set equipped with an order $\succ$ obeying transitivity and trichotomy (see page 18, if necessary). Show, by induction on the size $|A|$ of A, that A has a least element.*

We have thus seen that 'order numbers' of Peano, where one number succeeds another, can be used as 'counting numbers' where two finite sets have the same number of elements if and only if there is a bijection between them.

The process may be reversed so that we start with 'counting numbers' and end with 'order numbers'. To do this, we first choose particular finite sets to provide a scale of sizes.

Example 6.4.14 *(To be omitted by those who dislike hand-waving.) One way of providing scale sets, due to Von Neumann, is to look at $\mathbf{1} = \{\varnothing\}$ (the set with one element consisting of the empty set),*

$$\mathbf{2} = \{\varnothing, \mathbf{1}\} = \{\varnothing, \{\varnothing\}\}$$
$$\mathbf{3} = \{\varnothing, \mathbf{1}, \mathbf{2}\} = \{\varnothing, \{\varnothing\}, \{\varnothing, \{\varnothing\}\}\}$$
$$\vdots$$
$$\mathbf{n+1} = \{\varnothing, \mathbf{1}, \mathbf{2}, \ldots, \mathbf{n}\}.$$

If there is a bijection $f : \mathbf{n} \to E$, *we say that* E *is a set of the same size as* $\mathbf{n}$, *or, more informally, that* E *has n elements.*[5]

We can have a bijective function $f : A \to B$ even when A and B are infinite. Cantor used this idea to define *cardinal numbers* to measure the size of infinite sets. The following exercise provides a hint (but only a hint) as to how this might be done.

Exercise 6.4.15 *Let us call two non-empty*[6] *sets* A *and* B equipollent *if there is a bijection* $f : A \to B$.

 (i) *Let* $\mathcal{A}$ *be a collection (that is to say, a set) of non-empty sets. Show that equipollence is an equivalence relation on* $\mathcal{A}$.
 (ii) *Show, using earlier results in this section, that two non-empty finite sets are equipollent if and only if they have the same number of elements.*
(iii) *If* A *and* B *are non-empty sets, show that* $A \times B$ *is equipollent to* $B \times A$. *(We can think of this as the etherialisation of the diagram on page 10 which gave us* $3 \times 4 = 4 \times 3$.*)*
(iv) *Let* A *be the set of squares of the natural numbers* $\mathbb{N}^+$. *Show that* A *is equipollent to* $\mathbb{N}^+$ *although* A *is a subset of* $\mathbb{N}^+$ *which does not equal* $\mathbb{N}^+$. *(This result goes back to Galileo. In Theorem 8.2.7 we shall show that not all infinite sets are equipollent.)*

Cantor also generalised the Peano axioms by introducing 'transfinite induction' to obtain *ordinal numbers* (that is to say, ordered numbers) 'beyond' $\mathbb{N}^+$. (It is worth noting that his first use of this idea arose not in some deliberate attempt at generalisation, but in trying to solve a concrete problem in Fourier Analysis.) The new 'cardinal' and 'ordinal' arithmetics are very different from each other and from the familiar arithmetic developed in this book. But that is another story.

6.5 Objections

We have seen how to construct the strictly positive rationals from the natural numbers and the rationals from the strictly positive rationals. However, we did not construct the natural numbers, but, instead, presented a set of rules, the Peano axioms, which they obey, and shown how the other rules that we use

---

[5] (A very small print footnote.) At first sight, it may look as if we are getting the Peano axioms, and in particular the principle of induction, for free, but we need a set theoretic axiom called 'the axiom of infinity' (see page 49 in [12]) in order to talk about the set of Von Neumann 'numbers'.

[6] By convention, the empty set is equipollent to itself and not equipollent to a non-empty set.

follow from the axioms. In this section, we discuss various objections to this procedure.[7]

The Plain Man Objection

There exists a school of thought, ably represented by many journalists, English intellectuals and politicians, which says that anything can be explained in ten minutes. Holders of this view would consider the hundred pages or so that we have devoted to going from the Peano axioms to the rationals as airy-fairy mystification. Since nothing we have done affects the use of numbers for trade, building pyramids or tax collecting, they are right from their own point of view, and it would be foolish to argue with them.

However, any reader who has got this far must be someone who enjoys reasoning for its own sake. She will have enjoyed the proof that there are an infinity of primes and be intrigued by the following theorem of Dirichlet.

Theorem 6.5.1 [Dirichlet's theorem on primes in arithmetic progression]
If a and b are coprime strictly positive integers, then there are an infinity of primes of the form $an + b$ with n an integer.

Exercise 6.5.2

(i) *If a and b are strictly positive integers, but not coprime, explain why there can be at most one prime of the form $an + b$.*

(ii) *Use Exercises 5.5.10 and 5.5.11 to show that Dirichlet's theorem is true when $a = 4$.*

I think the reader will agree that such a remarkable theorem requires a cast-iron proof with every step examined thoroughly.[8] But every chain of reasoning must start from somewhere and that somewhere must be a collection of agreed principles, that is to say, axioms.

The Romantic Objection

A nineteenth-century mathematician might agree that arguments must start from common first principles, but object that we have an inherent understanding of the natural numbers and that we cannot reduce such an understanding to a set of axioms. If the romantic view is right, then, just as we stumbled across various properties of arithmetic in the first chapter of this book and then added the principle of induction later, so there may be other 'inherent properties' of the natural numbers not deducible from the Peano axioms, still to be discovered.

---

[7] For a serious philosophical treatment of these and related topics, I recommend my father's book [19].

[8] She will also not be surprised to be told that the only known proofs are quite hard.

Although, I think, few people now hold this view, it cannot be refuted. On the other hand, it is up to the romantic mathematicians to discover such an 'inherent property' and, until they do, we can continue to use the Peano axioms. (And, of course, even if such an 'inherent property' was discovered and the mathematical world agreed that the Peano axioms needed to be supplemented by a further axiom[9] we could still study the consequences of the Peano axioms by themselves.)

The Truth Seeker's Objection

Until the beginning of the nineteenth century, the axioms of Euclidean geometry were regarded as 'self evident'. It would be comforting if the Peano axioms were 'self evident', but the question then arises, 'self evident to whom?' There are many people, including, I must confess, the present author, to whom the existence of the natural numbers as described by the Peano axioms does appear evident. However, there are others, including some very clever mathematicians, for whom this is not the case. If something is not evident to a sufficient number of people it is not self evident.

If we look at mathematical practice, we do not see philosophers engaged in a quest for truth, but mathematicians seeking and checking deductions.[10] Pure mathematics viewed in this way is a game like chess or go. No one is forced to play chess, but, if you do play chess, you must make moves in accordance with the rules. No one is forced to study particular mathematical systems like the natural numbers or Euclidean geometry, but, once the axioms have been agreed, we must make our deductions in accordance with these axioms.

As Russell elegantly put it,

> Pure mathematics consists entirely of assertions to the effect that, if such and such a proposition is true of anything, then such and such another proposition is true of that thing. It is essential not to discuss whether the first proposition is really true, and not to mention what the anything is, of which it is supposed to be true. Both these points would belong to applied mathematics. We start, in pure mathematics, from certain rules of inference, by which we can infer that if one proposition is true, then so is some other proposition. These rules of inference constitute the major part of the principles of formal logic. We then take any hypothesis that seems amusing, and deduce its consequences. If our hypothesis is about anything, and not about some one or more particular things, then our deductions constitute

---

[9] This possibility is not excluded by Theorem 6.3.7, which states that all Peano systems are isomorphic *if we only consider successor properties*. Logicians have constructed explicit examples of statements about Peano systems which cannot be deduced from Peano's axioms and which do not contradict them. However, I do not know of any claim that these examples reveal 'inherent properties' of the natural numbers.

[10] A minor complication is introduced by the mathematician's habit of saying 'hence X is true' when she means 'hence X follows from our premises'.

mathematics. Thus mathematics may be defined as the subject in which we never know what we are talking about, nor whether what we are saying is true.

Russell, in Mathematics and the Metaphysicians[11] *[28]*

The Problem of Existence

When Glendower boasts[12]

I can call spirits from the mighty deep.

Hotspur replies

Why so can I, or so can any man
But will they come when you do call for them.

As Poincaré pointed out, laying out an axiomatic system like Peano's does not mean that a system of the type described by the axioms actually exists. I can describe a unicorn as a hoofed animal with a single horn, but the definition does not force a unicorn to exist.

One answer to this objection is that the axioms of set theory allow us to construct the natural numbers, but this is a bit like saying 'Since fairyland and all its inhabitants exist, it follows that unicorns exist.'

Another answer, which I favour, is that existence, like truth, is not a very useful concept when applied to complex mathematical systems. If we talk about the system $(\mathbb{Z}_2, +, \times)$ of integers modulo 2, first introduced on page 17, then even the most sceptical must admit that such a system exists. However, anyone who asks for a proof that an abstract infinite system like $\mathbb{N}^+$ exists she must be prepared to explain what sort of proof, if any, they would be prepared to accept.

Exercise 6.5.3 *Produce an upper estimate for the number of people who have lived since the invention of writing. Produce an upper estimate on the number of integers that someone can write down during her lifetime. Write down an integer n which (with very high probability) no one has ever written down before. In what sense did the number exist before you wrote it down? In what sense does it now exist?*

Suppose you now write down another such integer m. Why are you confident that $m \times n = n \times m$ although (with very high probability) no one has done the calculation?

You may care to recall the size of the numbers involved in our discussion of the Rabin code on page 102.

---

[11] Russell says 'The tone [of the article] is partly explained by the fact that the editor begged me to make the article "as romantic as possible".'

[12] Shakespeare, *Henry IV*, part I.

In my opinion, a reasonably clever and informed person has to be very clever and very brave to doubt the existence of very large integers. As his life shows, Yessenin-Volpin was very clever, very brave and doubted the existence of very large integers. When Friedman objected to this view

> ... he asked me to be more specific. I then proceeded to start with 2^1 and asked him whether this is real or something to that effect. He virtually immediately said yes. Then I asked about 2^2, and he again said yes, but with a perceptible delay. Then 2^3, and yes, but with more delay. This continued for a couple of more times, till it was obvious how he was handling this objection. Sure, he was prepared to always answer yes, but he was going to take 2^{100} times as long to answer yes to 2^{100} than he would to answering 2^1. [9]

Exercise 6.5.4 *Which of the following statements do you consider true? Why do you consider them true and in what sense?*

(i) *The Alan Turing featured in the science fiction novel* Cryptonomicon *actually existed.*

(ii) *The Alan Turing portrayed in the film* The Imitation Game *actually existed.*

(iii) *The Alan Turing described in the fine biography by Andrew Hodges actually existed.*

(iv) *The Alan Turing who wrote* On Computable Numbers, with an Application to the Entscheidungsproblem *actually existed.*

It should be noted that, although we consider existence to be an important property for physical objects like our next dinner, we are far less concerned about existence for more abstract things. I enjoy reading *The Pickwick Papers* although I do not believe that Mr Pickwick existed or, indeed, that anybody resembling Mr Pickwick ever existed.

The Problem of Consistency

Pure mathematicians, like the author and the mathematicians described by Russell, see mathematics as a kind of game. Provided we obey the rules of the game, it is meaningless (or, at least, unnecessary) to ask if the rules are 'true' or if the game 'exists'. Unfortunately, it does make sense to ask if the rules are consistent. If we can prove that some proposition is true and we can prove that the same proposition is false, then the standard logic used by mathematicians tells us that every proposition is true.[13]

---

[13] Traditionally, we say 'Ex falso quodlibet': you can deduce anything from a falsehood. Most mathematicians will be familiar with the way that what appears to be a great theorem turns out to be the consequence of a tiny error.

Informal proof of the statement just made. Suppose that the proposition P is true (where 'true' is shorthand for 'deducible from our axioms') and P is false (that is to say, the proposition 'not P' is deducible from our axioms). If Q is any proposition, we can prove it as follows:

Since P is true, the statement S, which says that at least one of P or Q is true, is itself true. Since P is false and the statement S is true, it follows that Q is true.   ∎

The game of proving theorems in a system where every theorem is automatically true is not a very interesting one. Thus the problem of consistency cannot be evaded in the way we have evaded the problem of truth and existence.[14]

One of the most important mathematical themes of the first third of the twentieth century was the search for a proof of the consistency of the Peano axioms and mathematics in general. In order to set about such a proof it is necessary to codify the informal rules of inference (for example, if P implies Q and Q implies R, then P implies R). This was not too hard, but was followed by a dreadful shock when Russell and others showed that standard mathematical modes of argument produced inconsistencies.

Example 6.5.5 [Russell's paradox] *Consider the set B of all sets E such that $E \notin E$. If $B \in B$, then $B \notin B$ and if $B \notin B$, then $B \in B$.*

These paradoxes showed the necessity for rules restricting the way that we define mathematical objects. The standard set of rules called the Zermelo–Fraenkel axioms tells us how we can construct sets (and so how we can define mathematical objects) and the rules of mathematical logic tell us how to recognise a mathematical statement. Once the task of producing these rules was completed, mathematicians could return to a search for consistency.

Surprisingly enough, it turned out that proofs of consistency could be given for quite interesting parts of mathematics such as the elementary Euclidean geometry of lines and circles.[15]

---

[14] Although most mathematicians would be happy to work with a consistent system without worrying about truth or existence, some would not. As Brouwer, one of the leaders of the minority view, put it, 'An incorrect theory which is not stopped by a contradiction is nonetheless incorrect, just as a criminal policy unchecked by a reprimanding court is nonetheless criminal' [31], page 336. It should also be pointed out that, if a system *exists*, it is automatically *consistent*.

[15] We have to be very careful not to introduce wider ideas when describing this system. Lines are simply things which intersect at points (otherwise we could introduce the real numbers by the back door), and though we can consider triangles and quadrilaterals we cannot talk about general n-gons since we do not have the Peano axioms.

However, in 1931, Gödel produced another tremendous shock by showing that we cannot prove the consistency of the kind of number theory[16] discussed in Chapter 4. Naturally, there was a certain amount of small print in the statement of his theorem, but the nature of his proof and that of a related theorem of Turing is so general that most mathematicians now believe that there is no acceptable way to show that a system of axioms rich enough to allow us to do standard mathematics is consistent.[17]

Since most mathematicians wish to do standard mathematics, they must do it without the comfort blanket of a proof of consistency. What cannot be cured must be endured, but the reader may be somewhat comforted by the thought that although there has been an active search for inconsistency in standard mathematics (that is to say, mathematics based on standard mathematical logic and Zermelo–Fraenkel set theory) for the past century, no inconsistency has been found.

If an inconsistency is found, then it could be one of two types. It is possible that it could be a 'technical problem' solved by tinkering with the rules governing the construction of sets.[18] Such an outcome would be the mathematical sensation of the decade.

Alternatively, the inconsistency could lie deep and force mathematicians to go against Hilbert's rallying cry 'No one shall expel us from the paradise that Cantor has created' by greatly restricting (or, in the worst case, abandoning) the use of infinite sets. Such an outcome would be the intellectual sensation of the century.

---

[16] In particular, the kind of number theory which permits the statement and proof of results like the existence and uniqueness of factorisation into primes.

[17] The word 'acceptable' involves some more small print. There are many non-technical accounts of Gödel's theorem. An old, but good, one was written by Nagel and Newman [24]. There is a good technical account in [16], a book which forms an excellent starting point for an ambitious student anxious to know about mathematical logic and advanced set theory. As Huckleberry Finn might say 'The statements was interesting, but tough.'

[18] Mathematicians have produced a variety of set theories. Conway likens the situation to a rowing boat labelled ZF towing a long line of similar boats. If ZF sinks, we simply transfer to another boat.

PART III

The Real Numbers (and the
Complex Numbers)

7

What Is the Problem?

7.1 Mathematics Becomes a Profession

Writing about the time around 1630, Rouse Ball speaks of it as being

> manifestly characterised by the feeling that mathematics should be studied for the sake of its applications to astronomy (including astrology[1] therein), navigation, mensuration and surveying; but it was tacitly assumed that even in these subjects its uses were limited and that a knowledge of it was in no way necessary to those who applied the rules deduced therefrom while it was generally held that its study did not form any part of a liberal education. [2]

The only example of mathematics studied for its own sake was Euclidean geometry and this was felt (essentially correctly) to be a completed subject in which only minor improvements could be made.

There was no such thing as a professional research mathematician, that is to say, someone who makes a living by finding new mathematics or novel applications of old mathematics.

In the two hundred years following the invention of the calculus, everything changed. When the Queen of Prussia asked Leibniz about Newton, he replied 'that, taking Mathematicks from the beginning of the world to the time of [Newton] what [Newton] had done was much the better half' [32]. Men like Leibniz and Newton were no longer content to interpret the ancients; they were determined to surpass them.

The new mathematics gave promise of valuable practical applications. The celestial mechanics of Newton and his successors converted the heavens into a clock accurate enough to allow the determination of longitude, though, in the end, terrestrial clock makers produced even more accurate clocks. Sometimes

---

[1] TWK's note. If we take astrology in the wider sense of the confident application of inappropriate mathematics to predict the future, then we still have plenty of astrologers. See, for example, [23].

the promise would take time to materialise. Static mechanics did revolutionise bridge design, but only after new materials like iron and steel were introduced. Sometimes, other ideas would prove superior. It could be said (and in England it often was said) that the French studied the theory of steam engines, but the English built them.[2] Often the promise would not materialise at all.[3] But when the promise was fulfilled, as in the design of fortifications and the pricing of insurance, the benefits were great.

From the time of the French Revolution, the French State supported a system whose main purpose was to produce mathematically trained engineers and school teachers, but which also produced a small cohort of professional mathematicians.[4] Similar developments took place in Prussia and other German states and, in time, these were echoed elsewhere.

As mathematics lecturers know, it is easy to give wrong or incomplete proofs in a lecture provided you are not aware that you are doing so. Once you are aware of the fallacy or the gap, lecturing becomes much harder. The reconstruction of calculus during the nineteenth century owes much to the needs of conscientious teachers within the new systems of advanced education. (Complaints about Cauchy's courses show that this conscientiousness was not necessarily appreciated by students or colleagues.)

7.2 Rogue Numbers

The everyday life of the farmer involves natural numbers. The everyday life of a money changer or a tax collector involves rationals. However, the abstract geometry of the Greeks revealed objects which looked like numbers but which were not (or might not be) rational numbers. The reader has met such objects in school when she was told to solve problems involving $\sqrt{2}$ and π. The Greeks knew that $\sqrt{2}$ was not rational[5] (see Theorem 4.4.12) and strongly suspected that π was not.[6] We have seen a proof that the equation

---

[2] However, Carnot's theory is applicable to everything from power stations to jet engines and hurricanes.

[3] There is a story of a meeting of a modern British research-grant-giving body whose representative boasted that 90% of its grants had a successful outcome. A senior member of his audience replied that such a high success rate was a disgrace.

[4] An example of a mathematical lineage dating back to the Revolution is given by Aline Bonami, whose thesis adviser was Yves Meyer, whose thesis adviser was Kahane, whose thesis adviser was Mandelbrojt, whose thesis adviser was Hadamard, whose thesis adviser was Picard, whose thesis adviser was Darboux, whose thesis adviser was Chasles, whose thesis adviser was Poisson, whose thesis adviser was Lagrange.

[5] That is to say, they were in possession of a geometric version of this statement which simple-minded people like the present author would consider equivalent.

[6] The second part of the sentence is even more ahistorical than the first. What the Greeks actually studied was the problem known as the squaring of the circle.

$$x^2 = 2$$

has no rational solution (see Theorem 4.4.12) and Lambert proved that 'π is irrational' in 1768.

Experience shows that even the most exceptional child is happy to treat these objects as numbers and accept without question such statements as

$$\pi + 3 = 3 + \pi \text{ and } 3 \times \pi = \pi \times 3.$$

Now that the reader is older, she should, perhaps, be more sceptical. When she was in infant school, the formula $2 + 3 = 3 + 2$ may have been illustrated by a pattern of dots

$$\bullet \ \bullet \ | \ \bullet \ \bullet \ \bullet = \bullet \ \bullet \ \bullet \ | \ \bullet \ \bullet,$$

as in our first chapter. When she was introduced to fractions, the formula

$$\frac{3}{4} + \frac{2}{5} = \frac{2}{5} + \frac{3}{4}$$

may have been proved by 'clearing the denominator' (here, by multiplying both sides by 20) to get

$$15 + 8 = 8 + 15,$$

which in turn may be replaced by a pattern of dots. What picture of dots permits us to intuit that $3 + \pi = \pi + 3$?

The reader may well reply that π and 3 should be thought of as lengths. We take a piece of string of length π and attach a piece of string of length 3 to it to obtain a string of length $3 + \pi$ which we place on a revolving table. Turning the table through half a revolution, we get a piece of string of length π to which is attached a piece of string of length 3. Of course, this now means that we have sheep numbers for counting with and string numbers for measuring with[7] and it is not entirely clear how they are to be connected.

Things are even less clear when we turn to multiplication. Our first thought might be to define $a \times b$ as the area of a rectangle of sides a and b, but this involves the new undefined notions of rectangles and areas. The reader may feel happy visualising

$$\sqrt{2} \times 3 = 3 \times \sqrt{2},$$

but an expression like

$$\sqrt{2} \times \pi^2 + 2 \times \sqrt{3} = \sqrt{2} \times (\pi^2 + \sqrt{6})$$

---

[7] Those philosophers who considered the problem used the word 'number' for what we have called 'sheep numbers' and 'quantity' for what we have called 'string numbers'. (See the quotation from Hume at the beginning of this book.)

looks to me like a geometric mess. There is a further *dimensional* problem when we look at the formula

$$l = 2\pi r,$$

connecting the length l of the circumference of a circle with its radius r. If we follow our naive approach to multiplication, then on the one side we have l, a length, and on the other side (considering 2π as a single number) an area or (considering 2 and π as distinct numbers) a volume.

It is believed that the first person to face and conquer these difficulties was Eudoxus of Cnidus.[8] His work is reported in Euclid's fifth book, *On Proportion* (that is to say, *On Ratio*; the title tells us how Eudoxus avoided dimensional problems). The fifth book of Euclid was widely held to be the most difficult part of the *Elements* and it is possible that most of the students who studied it and many of those who taught it failed to understand the problem it solved. The mathematicians like Dedekind who produced the first constructions of the real numbers $\mathbb{R}$ acknowledged that Eudoxus had solved similar problems in a similar manner. However, the modern treatment, as I hope to show, is inspired by a different question and goes rather deeper.

Our discussion needs the notion of an *ordered field*.

Definition 7.2.1 *An* ordered field $(\mathbb{F}, +, \times, >)$ *is a field* $(\mathbb{F}, +, \times)$ *(see Definition 5.1.1) together with an order $>$ having the following properties.*

  (x) If $a > b$ and $b > c$, then $a > c$. (Transitivity of order)
  (xi) Exactly one of the following conditions holds: $a > b$ or $b > a$ or $a = b$. (Trichotomy)
  (xii) If $a > b$, then $a + c > b + c$. (Order and addition)
  (xiii) If $a > b$ and $c > 0$, then $a \times c > b \times c$. (Order and multiplication)

Exercise 7.2.2 *Show that conditions (x) and $(xiii)$ can be replaced by the following conditions.*

  (x)' If $a > 0$ and $0 > b$, then $a > b$.
  (xiii)' If $a > 0$ and $c > 0$, then $a \times c > 0$.

Exercise 7.2.3 *Suppose that $(\mathbb{F}, +, \times, >)$ is an ordered field and $\mathbb{G}$ is a subfield of $(\mathbb{F}, +, \times)$ (see Definition 5.1.7). Show that $(\mathbb{G}, +, \times, >)$ is an ordered field with the inherited relation $>$ for order.*

[8] Also the first person to give a coherent account of planetary motion. His treatment, with many further additions, remained dominant until Copernicus.

More vividly, a sub-field of an ordered field is automatically an ordered subfield.

Exercise 7.2.4 *We work in an ordered field* $(\mathbb{F}, +, \times, >)$, *taking* $a, b \in \mathbb{F}$. *We define*

$$|a| = \begin{cases} a & \text{if } a \geq 0, \\ -a & \text{if } a < 0. \end{cases}$$

 (i) *Show that* $|a| = 0$ *if and only if* $a = 0$.
 (ii) *Show that* $|-a| = |a|$.
 (iii) *Show that* $|a \times b| = |a| \times |b|$.
 (iv) *Show that* $|a| + |b| \geq |a + b|$.
 (v) *Show that*

$$(a + b) + |a - b| = 2 \max\{a, b\}.$$

Find and prove a similar formula for $\min\{a, b\}$.

It is fairly easy to see that any ordered field contains a copy of $\mathbb{N}^+$.

Exercise 7.2.5 *Let* $(\mathbb{F}, +, \times, >)$ *be an ordered field with identity* **1** *and zero* **0**.

 (i) *Prove that* $\mathbf{1} > \mathbf{0}$.
 (ii) *Let* $f : \mathbb{N}^+ \to \mathbb{F}$ *be defined inductively by the conditions* $f(1) = \mathbf{1}$ *and* $f(n + 1) = f(n) + \mathbf{1}$. *Show, by induction on* n, *that* $f(m) > f(n)$ *whenever* $m > n$. *Why does this show that* f *is injective?*
 (iii) *Show that* $f(m + n) = f(m) + f(n)$ *and* $f(m \times n) = f(m) \times f(n)$ *for all* $n, m \in \mathbb{N}^+$.
 (iv) *Suppose that* $u : \mathbb{N}^+ \to \mathbb{F}$ *is an injective function which preserves* $>, +$ *and* $\times$. *By considering* $u(1) \times u(1)$, *or otherwise, show that* $u(1) = \mathbf{1}$. *Show, by induction, that* $u(n) = f(n)$ *for all* $n \in \mathbb{N}^+$.

Conclusion (iv) may be interpreted as saying that the copy of $\mathbb{N}^+$ is unique.

Exercise 7.2.6 *Give an example of a field which does not contain a copy of* $\mathbb{N}^+$. *Where does the argument of Exercise 7.2.5 break down?*

It is now easy to see that any ordered field contains a unique copy of $\mathbb{Q}$.

Exercise 7.2.7 *Let* $(\mathbb{F}, +, \times, >)$ *be an ordered field.*

*(i) Use Exercise 7.2.5 to show that there exists an injective function g :
$\mathbb{Q}^+ \to \mathbb{F}$ which preserves $+$, $\times$ and $>$. (Be careful to show that your
chosen g is well-defined.)*

(ii) Show that g is unique.

*(iii) Show that there exists an injective function $h : \mathbb{Q} \to \mathbb{F}$ which preserves
$+$, $\times$ and $>$. (Be careful to show that your chosen h is well defined.)*

(iv) Show that h is unique.

Exercise 7.2.7 sheds light on the various different ways in which we can
construct models of the rationals. In particular, it concludes the discussion at
the end of Section 3.4.

Exercise 7.2.8 *Show that an ordered field $(\mathbb{F}, +, \times, >)$ has no sub-fields apart
from itself if and only if it is isomorphic to $(\mathbb{Q}, +, \times, >)$.*

Thus, whenever we find a smallest extension of the natural numbers to an
ordered field, we obtain the same object (up to isomorphism). We follow our
standard procedure and write $\mathbb{Q}$ for the image of $\mathbb{Q}$ under h.

Exercise 7.2.9 *In Exercise 2.3.7 we showed, in effect, that there was a map f :
$\mathbb{N}^+ \to \mathbb{Q}^+$ which preserved $>$, $+$ and $\times$. Use the method of Exercise 7.2.5 (iv)
to show that it is unique.*

*What missing ingredient allows us to prove uniqueness now which we did
not have when we first looked at Exercise 2.3.7?*

State the analogous extension of Lemma 3.2.18.

So far $\mathbb{Q}$ is the only ordered field that we have seen, but it is easy to
manufacture others.

Exercise 7.2.10 *(The reader should follow the outline of this exercise, but
should not take it too seriously.) Consider $\mathbb{Q}[\sqrt{2}]$ with elements $\mathbf{a} = (a_1, a_2)$
where $a_1, a_2 \in \mathbb{Q}$. (Secretly $\mathbf{a} \underset{?}{=} a_1 + a_2\sqrt{2}$.) We equip $\mathbb{Q}[\sqrt{2}]$ with two
operations $\oplus$ and $\otimes$ defined by*

$$\mathbf{a} \oplus \mathbf{b} = (a_1 + b_1, a_2 + b_2)$$
$$\mathbf{a} \otimes \mathbf{b} = (a_1 b_1 + 2a_2 b_2, a_1 b_2 + a_2 b_1)$$

*and a relation $\ominus$ given by $\mathbf{a} \ominus \mathbf{b}$ if and only if at least one of the following three
conditions hold:*

(1) $a_1 - b_1 \geq 0$, $a_2 - b_2 \geq 0$ and $\mathbf{a} \neq \mathbf{b}$.

(2) $a_1 - b_1 \geq 0$, $(a_1 - b_1)^2 > 2(a_2 - b_2)^2$.

(3) $a_2 - b_2 \geq 0$, $2(a_2 - b_2)^2 > (a_1 - b_1)^2$.

(i) Show that, if $(a_1, a_2) \neq (0, 0)$, then

$$(a_1, -a_2) \otimes \left(\frac{a_1}{a_1^2 - 2a_2^2}, \frac{a_2}{a_1^2 - 2a_2^2} \right) = (1, 0).$$

(ii) Verify as many of conditions (i) to (ix) for a field as you feel you ought to.

(iii) Observe that $f : \mathbb{Q} \to \mathbb{Q}[\sqrt{2}]$ defined by $f(a) = (a, 0)$ is a bijection preserving addition, multiplication and order.

(iv) Show that the equation $\mathbf{a} \otimes \mathbf{a} = f(2)$ has a solution in $\mathbb{Q}[\sqrt{2}]$.

(v) Show that the equation $\mathbf{a} \otimes \mathbf{a} = f(3)$ has no solution in $\mathbb{Q}[\sqrt{2}]$.

(vi) Convince yourself informally that the order conditions hold, but that the proof is tedious. The reader should not waste time on writing things out, since once we have constructed $\mathbb{R}$, the result is immediate (see Exercise 8.1.20) by identifying $\mathbb{Q}[\sqrt{2}]$ with a sub-ordered-field of $\mathbb{R}$.

Exercise 7.2.10 shows that the statement 'We introduce the reals so that we can talk about the square root of two' is a bit simple-minded. We can accommodate rogue roots in the manner of Exercise 7.2.10. However, we suspect that not all rogue numbers are roots (a suspicion confirmed in Section 10.3) and that Exercise 7.2.10 is irrelevant to the more general problem.

Far more relevant is a little booklet published in 1585 by Simon Stevin first in Dutch and then in French as *La Disme*.

> What is it that is here propounded? Some wonderful invention? Hardly that, but a thing so simple that it scarce deserves the name invention .... But, just as a mariner who has found an unknown isle, may declare all its riches to the king, as, for instance, it having beautiful fruits, pleasant plains, precious minerals etc., without it being imputed to him as conceit; so I may speak freely of the great usefulness of this invention, a usefulness greater than I think any of you anticipates....
>
> The more important [the calculations of astronomy, navigation, land surveying, running a mint, and commerce] are, and the more laborious their execution, so much the greater is this discovery of decimal numbers which does away with all these difficulties. To speak briefly. La Disme teaches how all computations of the type of the four principles of arithmetic addition, subtraction, multiplication and division may be performed by whole numbers with as much ease as in counter-reckoning.
>
> Stevin La Disme, *translation from [29]*

The discovery, in which Stevin takes such justifiable pride, is the method of writing and calculating with decimal numbers. Stevin ends by advocating the decimalisation of measures of quantities, such as length and weight, and the decimalisation of currency. The first country to follow Stevin's advice was the revolutionary United States which, at the instigation of Jefferson, chose a decimal currency. The *dime* preserves the memory of *La Disme*. Stevin had predecessors (though he may not have known of them) and his notation was simplified and extended by his successors, but it is to his little booklet that we owe the introduction of decimals both to mathematics and to ordinary life. Si monumentum requiris, circumspice.[9]

The following exercises illustrate Stevin's remarks.

Exercise 7.2.11

(i) *The old British system of distance measurements had* 12 *inches to the foot,* 3 *feet to the yard,* 22 *yards to the chain,* 10 *chains to the furlong and* 8 *furlongs to the mile. Add the length* 5 *miles,* 6 *furlongs,* 6 *chains,* 12 *yards,* 2 *feet and* 3 *inches to the length* 7 *miles,* 6 *furlongs,* 2 *chains,* 14 *yards,* 2 *feet and* 7 *inches.*

(ii) *Add* 3065.731 *metres to* 9731.468 *metres.*

Exercise 7.2.12

(i) *Which is larger:* 3/19 *or* 4/23?

(ii) *Which is larger:* 0.353 *or* 0.361?

(iii) *Find the average of* 3/19, 4/23 *and* 4/21, *expressing the result as a fraction.*

(iv) *Find the average of* 0.353, 0.361 *and* 0.362 *correct to three places of decimals.*

Exercise 7.2.13

(i) *The old British system of currency dated back to Charlemagne who reintroduced and modified an ancient Roman system. There are* 4 *farthings in a penny (one pence),* 12 *pence in a shilling and* 20 *shillings in a pound.*[10]

The sum of 56 *pounds,* 12 *shillings,* 5 *pence and* 2 *farthings is lent out at a rate of* 3% *per year interest. If no money is paid back, what will be the total debt (in pounds, shilling, pence and farthings correct to the nearest farthing) after one year?*

[You may use a calculator, but my school days took place in the time of the old currency and before the 'hand calculator'. In real life such questions would be answered by using a booklet of tables.]

---

[9] If you seek his monument, look around you.

[10] There were also coins worth three pence, six pence, two shillings (called a florin), two shillings and six pence (called a half crown) and a ten shilling note. Lawyers' and doctors' bills used units of a guinea (twenty one shillings) – this was felt to be a gentlemanly way of disguising a 5% mark up.

(ii) *The US dollar is divided into* 100 *cents. The sum of* 56 *dollars and* 53 *cents is lent out at a rate of* 3% *per year interest. If no money is paid back, what will be the total debt (correct to the nearest cent) after one year?*

The first major use of Stevin's idea came with the invention of logarithms by Napier and their development by Briggs.

Example 7.2.14 *The logarithm to base* 10, $\log_{10}$ *is an injective function (whose existence we shall not prove in this book) defined for the strictly positive numbers with the property that* $\log_{10} a \times b = \log_{10} a + \log_{10} b$ *and* $\log_{10} 10 = 1$. *For three centuries it provided the fastest method of multiplying two numbers together. For example, to compute* 2.345×3.679, *we set* $a = 2.345$, $b = 3.679$. *We now consult a table of logarithms*[11] *and see that* $\log_{10} 2.345 = 0.3701$, $\log_{10} 3.679 = 0.5657$. *Adding gives*

$$\log(2.345 \times 3.679) = \log_{10} 2.345 + \log_{10} 3.679 = 0.3701 + 0.5657 = 0.9358.$$

Consulting our table of logarithms again, we see that

$$\log_{10} 8.626 = 0.9358$$

and we conclude that $2.345 \times 3.679 = 8.626$ *(to the appropriate accuracy*[12]*).*

Exercise 7.2.15 *Show how to divide using logarithms using as examples* $8.626 \div 3.679$ *and* $2.345 \div 3.679$.
[$\log_{10} 6.374 = 0.8044$.]

Exercise 7.2.16

(i) *If* $x > 0$ *and* n *is an integer, show that* $n \log_{10} x = \log_{10} x^n$.
(ii) *Let* a, b, u, v *be strictly positive integers. If*

$$\frac{a}{b} = \log_{10} \frac{u}{v}$$

show that $10^a v^b = u^b$. *If* u *and* v *are coprime, use uniqueness of factorisation (Theorem 4.4.3) to show that* $v = 1$ *and* $u = 10^k$ *for some strictly positive integer* k.
(iii) *Let* $x > 0$. *Show that* x *and* $\log_{10} x$ *will both be rational if and only if* $x = 10^n$ *for some integer* n.

---

[11] Notice, once again, the importance of printing, which allowed cheap and *accurate* reproduction of such tables.
[12] As with any calculation where we work with a fixed number of decimals, the accuracy degrades with the number of steps, but the degradation is slow and easy to control.

As Exercise 7.2.16 shows, even if a is rational, we will have $\log_{10} a$ rational only in very special cases. There is no reasonable way of marrying the use of logarithms with fractional representation and anybody who did serious calculations, whether astronomer, engineer, surveyor or ship's captain, was forced to use decimals.

Exercise 7.2.17 *Go to your nearest scientific library, which probably still has a dusty shelf dedicated to 'Mathematical Tables'. (Otherwise find a copy of logarithm tables on the internet, but that is not quite the same.) Take down a volume, open it and think of the work and ingenuity involved, not merely in computing the results by hand, but in typesetting and proof reading such a set of tables. Think too of the generation after generation for whom mastery of calculation using tables was the first step in any technical or scientific career.*

Wondrous is this foundation – the fates have broken
and shattered this city; the work of giants crumbles.
The roofs are ruined, the towers toppled,
frost in the mortar has broken the gate,
torn and worn and shorn by the storm,
eaten through with age. The earth's grasp
holds the builders,

The Ruin,[13] *translation by R. M. Liuzza*

Let us have an informal look at the decimal world based on what we learned at school. (There is a more formal treatment in Exercises 8.2.3 to 8.2.5.) It is a world very different from that we have discussed so far. When we write a number in decimal, it performs a dance of the seven veils, with each place of decimals revealing a little more about the number. When we write down $4/7$ as a fraction, we know exactly what it is, but the decimal notation just gives us a better and better idea of what it is, without ever giving us the number exactly[14]:

$$0.5, \ 0.57, \ 0.571, \ 0.5714, \ 0.57142, \ \ldots.$$

Because no number is ever fully revealed, decimal representation is democratic between rational numbers and the rest.

The reader will recall that a number is rational if and only if its decimal expansion is periodic from some place on. However large N is, we cannot tell from the first N places

---

[13] An Anglo–Saxon poem, possibly about the ruins of Roman Bath, found in the *Exeter Book*.
[14] This way of looking at things chimed in rather well with the rise of precision measurement in science and engineering. One generation measured things correct to the nearest inch, the next to the nearest tenths of an inch, their successors correct to the nearest hundredth and so on.

$$.a_1 a_2 \ldots a_N$$

of a decimal expansion if the number is rational (even if you detect periodicity, that periodicity may break down when you see the next place) or not (if you do not detect periodicity, this may be because the period is too long or because you have not yet reached the point where periodicity kicks in).

Speaking vaguely, periodicity is a rather special property and this suggests a picture in which the irrationals (that is to say, the non-rationals), instead of being the occasional rogue number, are in fact representative of the 'real numbers' and the rationals are the exceptions. Further, if we think a little more about the discussion of the last paragraph, we see that it shows that the rationals and irrationals are tightly intermingled, with every rational arbitrarily close to irrationals and every irrational arbitrarily close to rationals.

We shall not use the technology of decimal representation in what follows, but, like the nineteenth-century mathematicians who sought to make analysis rigorous, we shall keep the general decimal picture in mind.

7.3 How Can We Justify Calculus?

The invention of the calculus and its development by Newton, Leibniz, Euler and others produced an enormously powerful mathematical tool, but failed to provide a clear understanding of its underlying assumptions, When applied to physics, the guiding hand of nature allowed practitioners to avoid paradox. When practised for its own sake, it produced many beautiful formulae which could be admired without worrying too much about what they meant.[15]

Example 7.3.1

(i) $1 + \dfrac{1}{2} + \dfrac{1}{2^2} + \dfrac{1}{2^3} + \ldots \underset{?}{=} 2.$

(ii) $\dfrac{9}{10} + \dfrac{9}{10^2} + \dfrac{9}{10^3} + \ldots \underset{?}{=} 1.$ (*In decimal notation* $0.99999 \ldots \underset{?}{=} 1.$)

(iii) $1 + 2 + 2^2 + 2^3 + \ldots \underset{?}{=} -1.$ *Thus* $\infty \underset{?}{=} -1.$

Proof? (i) Let

$$S \underset{?}{=} 1 + \frac{1}{2} + \frac{1}{2^2} + \frac{1}{2^3} + \cdots .$$

---

[15] 'My mathematical tutors had never shown me any reason to suppose the Calculus anything but a tissue of fallacies' (Bertrand Russell in his *Autobiography*). Of course, the standards Russell demanded from others were unusually high.

Then, multiplying by 2,

$$2S \underset{?}{=} 2 + 1 + \frac{1}{2} + \frac{1}{2^2} + \frac{1}{2^3} + \cdots \underset{?}{=} 2 + S,$$

so $S \underset{?}{=} 2$.

(ii) Let

$$S \underset{?}{=} \frac{9}{10} + \frac{9}{10^2} + \frac{9}{10^3} + \cdots .$$

Then, multiplying by 10,

$$10S \underset{?}{=} 9 + \frac{9}{10} + \frac{9}{10^2} + \frac{9}{10^3} + \cdots \underset{?}{=} 9 + S,$$

so $9S \underset{?}{=} 9$ and $S \underset{?}{=} 1$.

(iii) Let

$$S \underset{?}{=} 1 + 2 + 4 + 8 + \cdots + 2^n + \cdots .$$

Then, multiplying by 2,

$$2S \underset{?}{=} 2 + 4 + 8 + \cdots \underset{?}{=} -1 + (1 + 2 + 4 + 8 + \cdots) \underset{?}{=} -1 + S$$

so $S \underset{?}{=} -1$. But $S \underset{?}{=} \infty$, so $\infty \underset{?}{=} -1$. ∎

Various proposals were put forward to provide foundations. Some mathematicians hoped to obtain certainty by grafting the calculus onto Greek geometry. This may have been one reason why Newton wrote his *Principia* in geometric language. Others, like Leibniz, may have hoped that, by using the same kind of trickery which we used in Exercise 7.2.10 to incorporate $\sqrt{2}$ into the number system (a better analogy is given by the construction of $\mathbb{C}$ in Section 9.1), they could produce 'infinitesimally small' elements and justify calculus that way. For a time, it was hoped that considering functions f as Taylor series, that is to say,

$$f(x) \underset{?}{=} a_0 + a_1 x + a_2 x^2 + a_3 x^3 + \cdots$$

would provide an answer.[16] It seems likely that most of those who thought about foundations underestimated the difficulties they faced.

---

[16] The influence of Newton meant that Cambridge mathematicians, in contrast to the rest of Europe, placed their hopes in geometry. They switched to Taylor series, just as the best European mathematicians began to realise that Taylor series did not provide a suitable foundation. 'An exploded theory or a disadvantageous practice, like a rebel or a patriot in distress, seeks refuge on our [English] shores to spend its last days in comfort if not in splendour' (Todhunter, *The Conflict of Studies*).

The decisive steps in providing firm foundations for the calculus were taken by Cauchy. Many of the definitions and results we shall need are in every first course on analysis.[17] The well-informed reader can do some judicious skipping.

Cauchy's most important decision was to base calculus on the notion of a limit.

Definition 7.3.2 *Let us work in an ordered field* $\mathbb{F}$. *We say that* $a_n \to a$ *(or, in words,* a_n *tends to a limit* a*) if, given* $\epsilon > 0$, *we can find an* N *such that* $|a - a_n| < \epsilon$ *for all* $n \geq N$.

Lemma 7.3.3 *We use the notations and assumptions of Definition 7.3.2.*

(i) If a_n *tends to a limit, then that limit is unique.*
(ii) If $a_n \to a$ *and* $b_n \to b$, *then* $a_n + b_n \to a + b$ *as* $n \to \infty$.
(iii) If $a_n \to a$ *and* $b_n \to b$, *then* $a_n b_n \to ab$ *as* $n \to \infty$.

Remark We shall follow the usual practice of writing $ab^{-1} = a/b$ and use the fact that any ordered field contains a unique copy of $\mathbb{Q}$, so that we may consider rationals like 2 and $1/2$ as members of $\mathbb{F}$.

Proof of Lemma 7.3.3. (i) Suppose that $a_n \to a$ and $b_n \to b$ as $n \to \infty$. If $a \neq b$, set $\epsilon = |a - b|/4$. Since $\epsilon > 0$, we can find N' and N'' such that $|a_n - a| < \epsilon$ for $n \geq N'$ and $|b_n - b| < \epsilon$ for $n \geq N''$. If we take $N = \max\{N', N''\}$, then

$$|a - b| = \big|(a - a_N) + (a_N - b)\big| \leq |a - a_N| + |a_N - b| < 2\epsilon = |a - b|/2,$$

which is impossible. The result follows by reductio ad absurdum.[18]

(ii) Let $\epsilon > 0$. We can find N' and N'' such that $|a_n - a| < \epsilon/2$ for $n \geq N'$ and $|b_n - b| < \epsilon/2$ for $n \geq N''$. If we take $N = \max\{N', N''\}$, then

$$\big|(a_n + b_n) - (a + b)\big) = \big|(a_n - a) - (b_n - b)\big| \leq |a_n - a| + |b_n - b|$$
$$< \epsilon/2 + \epsilon/2 = \epsilon$$

for all $n \geq N$.

(iii) Let $\epsilon > 0$. We can find N' and N'' such that

$$|a_n - a| < \min\left\{1, \frac{\epsilon}{2(|b| + 1)}\right\}$$

---

[17] One of the problems when teaching beginning analysis in France is that, quite properly, almost every theorem is called Cauchy's theorem.

[18] If we prove A is true by showing that the assumption A is not true leads to a contradiction, then the argument is called *reductio ad absurdum*.

for $n \geq N'$ and

$$|b_n - b| < \frac{\epsilon}{2(|a| + 1)}$$

for $n \geq N''$. If we take $N = \max\{N', N''\}$, then

$$|a_n b_n - ab| = |a_n(b_n - b) + b(a_n - a)| \leq |a_n(b_n - b)| + |b(a_n - a)|$$

$$= |a_n||b_n - b| + |b||a_n - a| \leq (|a_n - a| + |a|)|b_n - b| + |b||a_n - a|$$

$$\leq (1 + |a|)|b_n - b| + |b||a_n - a| < (1 + |a|)\frac{\epsilon}{2(|a| + 1)}$$

$$+ |b|\frac{\epsilon}{2(|b| + 1)} < \epsilon$$

for all $n \geq N$.   ∎

Exercise 7.3.4 *We use the notation and assumptions of Definition 7.3.2. Show that the following results are true.*

(i) *If $a_n = a$ for all n, then $a_n \to a$ as $n \to \infty$.*

(ii) *If $a_n \to a$, then $-a_n \to -a$ as $n \to \infty$.*

(iii) *If $a_n \to a$, $b_n \to b$ as $n \to \infty$ and there exists an M such that $b_m \geq a_m$ for all $m \geq M$, then $b \geq a$.*

(iv) *(Special cases of part (iii).) If $b_n \to b$ and $b_n \geq a$ for all n, then $b \geq a$. If $a_n \to a$ and $b \geq a_n$ for all n, then $b \geq a$.*

We now make precise our notion of a continuous function.

Definition 7.3.5 *Let $\mathbb{F}$ be an ordered field and let $f : \mathbb{F} \to \mathbb{F}$ be a function. We say that f is* continuous at $a \in \mathbb{F}$ *if, whenever $a_n \to a$, we have $f(a_n) \to f(a)$. We say that f is* continuous *if it is continuous at every point a of $\mathbb{F}$.*

Exercise 7.3.6 *Use Lemma 7.3.3 to show that, if $f, g : \mathbb{F} \to \mathbb{F}$ are continuous functions, then the sum function $h = f + g$, defined by $h(t) = f(t) + g(t)$ for all $t \in \mathbb{F}$, and the product function $k = f \times g$, defined by $k(t) = f(t) \times g(t)$ for all $t \in \mathbb{F}$, are continuous.*

It is a futile business to try and show that our definition of continuity corresponds to non-mathematical usage (as in 'They were awakened by a continuous hammering') because it does not.[19]

[19] Some indication of the cloud of confusion that surrounded the notion of continuity in Paris during the eighteenth century and in Oxford during the nineteenth is given by the definition of 'continuous function' which still appeared in the Oxford English Dictionary in 2018: 'a function that varies continuously, and whose differential coefficient therefore never becomes

7.4 The Fundamental Axiom of Analysis

The object of this part of the book is to provide suitable foundations for the calculus. The reader may be surprised not to find any mention of derivatives or integrals. However, the decision to choose continuity as fundamental is part of the Cauchy programme. (If the reader looks at a modern calculus text like [30], she will find that the first major theorems deal with continuous functions.)

We start with a property that we would wish every continuous function to have.

Definition 7.4.1 *Let* $\mathbb{F}$ *be an ordered field. We say that a function* $f : \mathbb{F} \to \mathbb{F}$ *has the* intermediate value property *if, whenever* $a < b$ *and* $f(a) \le 0 \le f(b)$, *we can find a* c *with* $a \le c \le b$ *and* $f(c) = 0$.

We shall show that every continuous function has the intermediate value property if and only if $\mathbb{F}$ obeys the *fundamental axiom of analysis*.

Definition 7.4.2 *Let* $\mathbb{F}$ *be an ordered field. We say that* $\mathbb{F}$ *satisfies* the fundamental axiom of analysis *if every increasing sequence bounded above tends to limit, that is to say, if* $u_n \in \mathbb{F}$, $u_n \le u_{n+1}$ *for all* $n \in \mathbb{N}^+$ *and there exists a* $U \in \mathbb{F}$ *such that* $u_n \le U$ *for all* $n \in \mathbb{N}^+$, *then there exists a* $u \in \mathbb{F}$ *such that* $u_n \to u$ *as* $n \to \infty$.

We first prove the 'only if' part of our statement.

Theorem 7.4.3 *Let* $\mathbb{F}$ *be an ordered field. If every continuous function* $f : \mathbb{F} \to \mathbb{F}$ *has the intermediate value property, then* $\mathbb{F}$ *satisfies the fundamental axiom.*

Proof. Suppose that $u_n \in \mathbb{F}$, $u_n \le u_{n+1}$ for all $n \in \mathbb{N}^+$ and there exists a U such that $u_n \le U$, yet u_n does not tend to limit. We show how to construct a continuous function $f : \mathbb{F} \to \mathbb{F}$ which does not have the intermediate value property.

Explicitly, we set

$$f(x) = \begin{cases} -1 & \text{if there exists an } m \text{ with } x < u_m, \\ 1 & \text{otherwise.} \end{cases}$$

infinite.' If this 'definition' meant anything, it would say that the cube root function given by $f(x) = x^{1/3}$ was not continuous at 0.

We observe that f is well defined, that $f(u_1) = -1$, that $f(U + 1) = 1$ and that f never takes the value 0. Thus f does not have the intermediate value property.

We need to show that f is continuous at every point t. More precisely, we need to show that, if $t_n \to t$, then $f(t_n) \to f(t)$ as $n \to \infty$. There are two cases to consider.

Case 1 Suppose that $t < u_m$ for some m and $t_n \to t$. Taking $\epsilon = u_m - t$, we know, from the definition of a limit, that we can find an N such that $|t_n - t| < u_m - t$ for all $n \geq N$. Automatically, $t_n < u_m$ and so $f(t_n) = -1$ for all $n \geq N$. Thus $f(t_n) \to -1 = f(t)$ as $n \to \infty$ and f is continuous at t.

Case 2 Suppose $t \geq u_m$ for all m and $t_n \to t$. We claim that there exists an N such that $t_n \geq u_m$ for all $n \geq N$ and all m. We argue by contradiction.

Suppose our claim is false, that is to say, that, given any N, we can find an $n' \geq N$ and an M such that $u_M > t_{n'}$. Since $t_n \to t$, we know that, given any $\epsilon > 0$, we can find an N such that $|t_n - t| < \epsilon$ for all $n \geq N$. We can now find an $n' \geq N$ and an M such that $t_{n'} < u_M$. Automatically

$$t_{n'} < u_M \leq u_m \leq t$$

and so

$$|t - u_m| \leq |t - t_{n'}| < \epsilon$$

for all $m \geq M$. Since ϵ can be chosen freely (at the cost of changing M), we see that $u_m \to t$, as $m \to \infty$ contrary to our initial assumption that the sequence does not tend to a limit.

Thus, by reductio ad absurdum, there exists an N such that $t_n \geq u_m$ for all $n \geq N$ and all m. It follows that $f(t_n) = 1$ for all $n \geq N$. Thus $f(t_n) \to 1 = f(t)$ as $n \to \infty$ and f is continuous at t.   ∎

Our proof of the converse of Theorem 7.4.3 uses ideas which are of great interest in themselves. Recall that any ordered field contains a unique copy of the rationals $\mathbb{Q}$ which we identify with $\mathbb{Q}$ itself.

Definition 7.4.4 [Axiom of Archimedes][20] *We say that ordered field* $\mathbb{F}$ *satisfies the axiom of Archimedes if* $1/n \to 0$ *as* $n \to \infty$.

Exercise 7.4.5 *Let* $\mathbb{F}$ *be an ordered field. Show that* $\mathbb{F}$ *satisfies the axiom of Archimedes if and only if, given* a, $b \in \mathbb{F}$ *with* $a > 0$, *we can find an* $n \in \mathbb{N}^+$ *such that* $na > b$.

---

[20] In spite of its name, the axiom was ascribed by Archimedes to his great predecessor Eudoxus (see page 140).

[*This statement is closer to the way that Eudoxus and Archimedes presented the axiom.*]

Exercise 7.4.6 *Let* $\mathbb{F}$ *be an ordered field which satisfies the axiom of Archimedes. Show that if* $c \in \mathbb{F}$ *we can find integers* n *and* m *such that* $m \leq c \leq n$.

We need a simple observation.

Exercise 7.4.7 *Let* $\mathbb{F}$ *be an ordered field. By considering* $-a_n$, *show that the fundamental axiom is equivalent to the statement that, if we have a decreasing sequence[21]* (a_n) *bounded below, then* a_n *tends to a limit.*

Theorem 7.4.8 *An ordered field* $\mathbb{F}$ *which satisfies the fundamental axiom must satisfy the axiom of Archimedes.*

We have no difficulty in thinking of ordered fields like $\mathbb{Q}$ for which the axiom of Archimedes holds. The reader may experience greater difficulty in thinking of an ordered field for which it does not, but we shall give an example in Section 10.4. The existence of such examples shows that the proof of Theorem 7.4.8 cannot be entirely trivial.

Proof of Theorem 7.4.8. Consider the sequence $(1/n)$. This is a decreasing sequence bounded below by 0 and so must tend to a limit α. Thus, given any $\epsilon > 0$, we can find an N such that $|\alpha - 1/n| < \epsilon$ for all $n \geq N$ and so, automatically, $|\alpha - 1/n^2| < \epsilon$ for all $n \geq N$. Thus $1/n^2 \to \alpha$ as $n \to \infty$.

However, taking $a_n = b_n = 1/n$ in Lemma 7.3.3 (iii), we see that

$$\frac{1}{n^2} = \frac{1}{n} \times \frac{1}{n} \to \alpha \times \alpha = \alpha^2,$$

as $n \to \infty$, so the uniqueness of limits (Lemma 7.3.3 (i)) tells us that

$$\alpha^2 = \alpha.$$

Thus $\alpha = 0$ or $\alpha = 1$. Since $1/2 \geq 1/n$ for all $n \geq 2$, Exercise 7.3.4 (i) tells us that $1/2 \geq \alpha$ and so $\alpha = 0$. ∎

Exercise 7.4.9 *We work in an ordered field* $\mathbb{F}$ *which satisfies the fundamental axiom. Show, by imitating the proof of Theorem 7.4.8, that, if* $1 > x \geq 0$, *we have* $x^n \to 0$ *as* $n \to \infty$. *Deduce that, if* $|x| < 1$, *then* $x^n \to 0$ *as* $n \to \infty$.

---

[21] We write (a_n) to mean the infinite sequence $a_1, a_2, a_3, \dots$.

We shall look at an important consequence of the axiom of Archimedes in Theorem 7.4.11, but we first prove the promised converse of Theorem 7.4.3.

Theorem 7.4.10 *Let $\mathbb{F}$ be an ordered field which satisfies the fundamental axiom of analysis. Then every continuous function $f : \mathbb{F} \to \mathbb{F}$ satisfies the intermediate value condition.*

This may be considered the first major theorem of modern analysis. It was proved independently by Bolzano and Cauchy using a method not far removed from the one we shall use.

Proof. Suppose that $f : \mathbb{F} \to \mathbb{F}$ is continuous, that $a < b$ and that $f(a) \le 0 \le f(b)$.

We make an inductive definition as follows. We establish our base case by setting $a_0 = a$ and $b_0 = b$ and observing that $a_0 < b_0$, $f(a_0) \le 0 \le f(b_0)$. We now suppose that we have found $a_n < b_n$ such that $f(a_n) \le 0 \le f(b_n)$. We set $c_n = (a_n + b_n)/2$ (informally, c_n is the mid-point of a_n and b_n). If $f(c_n) \le 0$ we set $a_{n+1} = c_n$, $b_{n+1} = b_n$. If $f(c_n) > 0$ we set $a_{n+1} = a_n$, $b_{n+1} = c_n$. In either case we now have

(i)$_n$ $a_n \le a_{n+1} < b_{n+1} \le b_n$,
(ii)$_n$ $f(a_{n+1}) \le 0 \le f(b_{n+1})$,
(iii)$_n$ $b_{n+1} - a_{n+1} = (b_n - a_n)/2$.

Conditions (i)$_n$ and (ii)$_n$ complete the inductive step.

A simple induction shows that

$$a = a_0 \le a_n \le a_{n+1} < b_{n+1} \le b_n \le b,$$

so the a_n form an increasing sequence bounded above by b. The fundamental axiom now shows that $a_n \to c$ for some c. Since $a \le a_n \le b$, we have $a \le c \le b$. Since f is continuous, $f(a_n) \to f(c)$ and, since $f(a_n) \le 0$, we have $f(c) \le 0$. [Notice the repeated use of Exercise 7.3.4 (iv).]

We now consider the behaviour of b_n. Another simple induction shows that

$$b_n - a_n = (b_0 - a_0)2^{-n}$$

for all n. The axiom of Archimedes shows that $2^{-n} \to 0$ and so $b_n - a_n \to 0$, whence

$$b_n = (b_n - a_n) + a_n \to 0 + c = c$$

as $n \to \infty$. Since f is continuous, $f(b_n) \to f(c)$ and, since $f(b_n) \ge 0$, we have $f(c) \ge 0$. The previous paragraph showed us that $f(c) \le 0$, so $f(c) = 0$. ∎

The method of proof we used above is called successive bisection or lion hunting (at every stage, we halve the size of an interval which we believe contains a lion[22]).

The axiom of Archimedes can be expressed in another way which readers who have done a first course in analysis may find more familiar.

Theorem 7.4.11 *Let $\mathbb{F}$ be an ordered field. The following statements are equivalent.*

(i) $\mathbb{F}$ satisfies the axiom of Archimedes.
(ii) If $x \in \mathbb{F}$ and $\epsilon > 0$, then we can find $u \in \mathbb{Q}$ such that $|x - u| < \epsilon$.

We shall not use the term, but the customary statement of (ii) is that '$\mathbb{Q}$ is *dense* in $\mathbb{F}$'.

Proof of Theorem 7.4.11. We first show that (ii) implies (i). Suppose that (ii) is true and we are given $\epsilon' > 0$. Since $\epsilon'/2 \in \mathbb{F}$ and $\epsilon'/4 > 0$, setting $x = \epsilon'/2$ in (ii) tells us that there is a $u \in \mathbb{Q}$ such that $|u - \epsilon'/2| < \epsilon'/4$. Automatically $0 < u < \epsilon'$. Since u is a strictly positive rational, we may write $u = p/q$ where p and q are strictly positive integers. Automatically

$$0 < \frac{1}{q} \le \frac{p}{q} = u < \epsilon'$$

and so $0 < 1/n < \epsilon'$ for all $n \ge q$. The axiom of Archimedes follows.

We now show that (i) implies (ii). Suppose that $\mathbb{F}$ satisfies the axiom of Archimedes, that $x \in \mathbb{F}$ and $\epsilon > 0$. We wish to show that there is a $u \in \mathbb{Q}$ such that $|x - u| < \epsilon$. Since $|(-x) - (-u)| = |x - u|$, we may assume that $x \ge 0$. If $x = 0$, we may take $u = 0$. Thus we may suppose $x > 0$.

By the axiom of Archimedes, we can find a strictly positive integer n such that $1/n < \epsilon$. We now look at the set E of non-negative integers r such that $r/n \le x$. Since $0 \in E$ the set is non-empty. However, the axiom of Archimedes tells us that there is a strictly positive integer N such that $1/N < 1/(nx)$ and so $x < N/n$. Automatically, all the members e of E must satisfy $e < N$. Thus E is bounded and we know (see Lemma 4.1.8) that a bounded non-empty set of integers has a largest element m, say. We now have $m/n \le x < (m + 1)/n$ and so

$$\left| x - \frac{m}{n} \right| < \frac{1}{n} < \epsilon.$$

Setting $u = m/n$, we have the desired result. ∎

---

[22] See *Method 4* of the paper entitled *A mathematical contribution to big game hunting* by H. Pétard which appeared in the American Mathematical Monthly 1938 and is reprinted in [4].

We use the intermediate value theorem (Theorem 7.4.10) to show that every non-negative number in an ordered field obeying the fundamental axiom has a square root. (Contrast Theorem 4.4.12.)

Lemma 7.4.12 *If we work in an ordered field* $\mathbb{F}$*, obeying the fundamental axiom, then, if* $a \geq 0$*, the equation* $y^2 = a$ *has unique solution with* $y \geq 0$*.*

Proof. Let $f : \mathbb{R} \to \mathbb{R}$ be given by $f(t) = t^2 - a$. By Exercise 7.3.6, f is continuous.

Since $f(0) = -a \leq 0 \leq (1 + a)^2 - a = f(a + 1)$, the intermediate value theorem tells us that there exists a y with $0 \leq y \leq a + 1$ such that $f(y) = 0$, and so $y^2 = a$. We remark that, if $t^2 = a$ and $t \neq y$, then

$$0 = t^2 - y^2 = (t - y)(t + y)$$

with $t - y \neq 0$ so, since $\mathbb{F}$ is a field, $t + y = 0$ and $t = -y$. ∎

As is standard, we write $\sqrt{a}$ for the non-negative solution of the equation $y^2 = a$ with $a \geq 0$.

Exercise 7.4.13 *We continue with the notation and assumptions of Lemma 7.4.12. If* $a > b > 0$*, show that* $\sqrt{a} > \sqrt{b}$*.*

Exercise 7.4.14 *We work in an ordered field* $\mathbb{F}$ *obeying the fundamental axiom. Throughout, n is a strictly positive integer.*

(i) *Show, by induction, that the function* $f_n : \mathbb{F} \to \mathbb{F}$ *defined by* $f_n(x) = x^n$ *is continuous. Show also that* $f_n(y) > f_n(x) > 0$ *whenever* $y > x > 0$*.*

(ii) *Show that, if* $a > 0$*, the equation* $x^n = a$ *has a solution.*

(iii) *Let* $a > 0$*. For which values of n does the equation* $x^n = -a$ *have a solution? For which values of n does the equation* $x^n = a$ *have a unique solution? Give reasons.*

7.5 Dependent Choice

From time to time in the work that follows we shall use arguments of the form 'choose an appropriate x_1. Once x_n has been chosen appropriately, choose an appropriate x_{n+1} and so on. In this way we obtain a sequence of appropriate x_j for all $j \in \mathbb{N}^+$.' At first sight this looks as though we are using definition by induction, but, looking at Theorem 4.2.1, we see that we have replaced a well-defined procedure for finding x_{n+1}, given x_n, by the words 'chosen appropriately'.

Notice that, having defined $n!$ by induction, I am confident that, provided we were prepared to do the work and careful to avoid mistakes, the reader and I would obtain the same value for 20! or 100!. However, if we were each to write an essay on the factorial function, then, even if we were to use the same first sentence and choose our words 'appropriately', it is very probable that my twentieth sentence would bear little or no relation to her twentieth sentence.[23]

It turns out that, while there is no problem in making a finite number of choices, we need to add a specific extra rule of reasoning to allow an infinite sequence of choices.

Axiom of dependent choice *Let X be a non-empty set and S a set of ordered pairs (g, h) with g, $h \in X$. Suppose that whenever $g \in X$ there exists an $h \in X$ with $(g, h) \in S$. Then if $x_1 \in S$ there exists a sequence (x_n) with $(x_n, x_{n+1}) \in S$ for all $n \geq 1$.*

Remark It may be helpful to think of S as the set of ordered pairs (g, h), where h is a permitted successor of g.

The axiom of dependent choice has the following useful consequence known as 'countable choice'.

Theorem 7.5.1 *Let E_1, E_2, ... be non-empty sets. Then there exists a sequence (y_n) with $y_n \in E_n$ for all $n \geq 1$.*

Proof. Take X to be the collection of ordered pairs $\underline{x} = (u, n)$ with $u \in E_n$ and $n \in \mathbb{N}^+$. Let S be the collection of ordered pairs $\big((v, n), (u, n + 1)\big)$ such that (v, n), $(u, n + 1) \in X$ (so we have an ordered pair of ordered pairs).

If $\underline{g} = (v, n) \in X$, then since E_{n+1} is non-empty there exists a $u \in E_{n+1}$ and taking $\underline{h} = (u, n + 1)$ we have $(\underline{g}, \underline{h}) \in S$.

Since E_1 is non-empty, there exists a $y_1 \in E_1$. If we take $\underline{x}_1 = (y_1, 1)$, the axiom of dependent choice tells us that there exists a sequence $(\underline{x}_n)$ in X with $(\underline{x}_n, \underline{x}_{n+1}) \in S$ for all $n \geq 1$. By induction, $\underline{x}_n = (y_n, n)$ so $y_n \in E_n$ and we have a sequence (y_n) of the required type. ∎

Russell used to illustrate this result by considering a millionaire who owns a chest of drawers with an infinite number of drawers labelled $1, 2, 3, \ldots$. Each drawer contains a pair of shoes and a pair of socks. If he wants one shoe from each drawer, then he can always choose the left shoe, but, if he wants one sock from each drawer, he needs countable choice.[24]

---

[23] This idea is played with in Borge's *Pierre Menard, Author of the Quixote*, a story which should also be considered when comparing this book with [20].

[24] See, for example, *My Philosophical Development* (Bertrand Russell, page 77).

It is often the case that, by using a little thought, we can replace a use of the axiom of dependent choice by a more complicated inductive definition along the lines of Theorem 4.2.1 (think of the shoes in Russell's example). However, logicians have shown that several central theorems of analysis require us to use the axiom of dependent choice.

A good example of the use of countable choice is given by the proof of the following theorem which we shall not use, but which appears in some form in almost every first course of analysis. It shows that, provided we accept countable choice, if we work over an Archimedean field (that is to say, a field satisfying the axiom of Archimedes), we may think of continuous functions as 'machines such that sufficiently small changes in inputs produce small changes in outputs'.

Theorem 7.5.2 *Let $\mathbb{F}$ be an ordered field satisfying the axiom of Archimedes and let $f : \mathbb{F} \to \mathbb{F}$ be a function. The following two statements are equivalent.*

(i) *f is continuous at a (that is to say, $f(x_n) \to f(a)$ whenever $x_n \to a$ as $n \to \infty$).*

(ii) *Given any $\epsilon > 0$, there exists a $\delta > 0$ (depending on a and ϵ) such that $|f(a) - f(b)| < \epsilon$ whenever $|a - b| < \delta$.*

Proof. The proof that (ii) implies (i) is straightforward. Suppose (ii) is true and $a_n \to a$. We know that, given $\epsilon > 0$, we can find a $\delta > 0$ such that $|f(a) - f(b)| < \epsilon$ whenever $|a - b| < \delta$. We also know that we can find an N such that $|a_n - a| < \delta$ for all $n \geq N$ and so

$$|f(a) - f(a_n)| < \epsilon$$

for all $n \geq N$. Thus (ii) implies (i).

Our proof that (i) implies (ii) uses Theorem 7.5.1. Suppose (ii) is false. Then we can find an $\epsilon > 0$ such that, given any $\delta > 0$, we can find a b such that $|a - b| < \delta$, but $|f(a) - f(b)| \geq \epsilon$. In particular, we can find a_n such that $|a - a_n| < 1/n$, but $|f(a) - f(a_n)| \geq \epsilon$. Thus $a_n \to a$, but $f(a_n) \not\to a$, so (i) is false. Thus (i) implies (ii). ∎

Of course, the fact that we have used countable choice in our proof does not tell us that countable choice is required, but logicians have shown that a version of countable choice is indeed needed.

If the reader feels that the axiom of dependent choice is a natural rule of reasoning, she is in good company. Most mathematicians accept it without hesitation and many of the remainder accept it because they believe that it produces richer (or, at least, more convenient) mathematical systems. Gödel showed that, if the Zermelo–Fraenkel system is consistent, then adding the axiom of dependent choice (or indeed the much stronger 'axiom of choice') will not make it inconsistent.

We now return to the study of the fundamental axiom.

7.6 Equivalent Forms of the Fundamental Axiom

In this section we collect some equivalent forms of the fundamental axiom. This is not a simple academic exercise. Each equivalent form is a powerful tool in its own right and each suggests different ways of generalising analysis to systems which are not ordered fields.

Because of the importance of these equivalent forms, most first courses in analysis contain an extensive discussion of them. The reader will thus be meeting ideas which she has met before or can expect to meet in the future and should feel free to run through the section quite quickly.

We start with the Bolzano–Weierstrass property.

Definition 7.6.1 *Let $\mathbb{F}$ be an ordered field. We say that $\mathbb{F}$ has the Bolzano–Weierstrass property if, whenever we have a sequence (x_j) with $x_j \in \mathbb{F}$ such that there exists an $A \in \mathbb{F}$ with $|x_j| \leq A$ for all $j \geq 1$, we can find $1 \leq j(1) < j(2) < \cdots$ and $x \in \mathbb{F}$ such that $x_{j(n)} \to x$ as $n \to \infty$.*

In other words, an ordered field has the Bolzano–Weierstrass property if every bounded sequence has a convergent subsequence.

Theorem 7.6.2 *Let $\mathbb{F}$ be an ordered field. The following two statements are equivalent.*

(i) $\mathbb{F}$ has the Bolzano–Weierstrass property.
(ii) $\mathbb{F}$ obeys the fundamental axiom of analysis.

Proof. We split the proof into two parts.

The Bolzano–Weierstrass property implies the fundamental axiom.
Suppose that $\mathbb{F}$ satisfies (i) and (a_j) is an increasing sequence bounded above. By the Bolzano–Weierstrass property, we can find $1 \leq j(1) < j(2) < \cdots$ and $a \in \mathbb{F}$ such that $a_{j(n)} \to a$ as $n \to \infty$. We claim that $a_n \to a$ as $n \to \infty$.

We observe that, since the $a_{j(n)}$ form an increasing sequence, $a_{j(n)} \leq a$ for each n. Given $\epsilon > 0$, we know that we can find an N such that $|a_j(n) - a| < \epsilon$ for $n \geq N$ and so $a - \epsilon < a_{j(n)} \leq a$ for $n \geq N$. Now suppose $m \geq j(N)$. We can find an $M > N$ such that $j(M) > m \geq j(N)$, so

$$a - \epsilon < a_{j(N)} \le a_m \le a_{j(M)} \le a$$

and $|a - a_m| < \epsilon$. Thus $a_m \to a$ as $m \to \infty$.

The fundamental axiom implies the Bolzano–Weierstrass property.
Suppose that $\mathbb{F}$ satisfies (i) and (a_j) is a sequence such that $|a_j| \le A$ for all j.

Let us call a strictly positive integer j 'far-seeing' if $a_j \ge a_r$ for all $r \ge j$. Either the set of far-seeing integers is finite or it is not. If there are only finitely many far-seeing integers, then there exists an N such that no j with $j \ge N$ is far-seeing. We now define $j(1) = N$ and, once $j(n)$ is defined, we let $j(n+1)$ be the least integer $r > j(n)$ such that $a_r > a_{j(n)}$. This inductive definition gives an increasing sequence $(a_{j(n)})$ bounded above by A and the fundamental axiom tells us that there exists an a with $a_{j(n)} \to a$ as $n \to \infty$.

If there are infinitely many far-seeing integers, we can list them as $m(1) < m(2) < m(3) < \cdots$. By definition,

$$a_{m(1)} \ge a_{m(2)} \ge a_{m(3)} \ge \cdots ,$$

so the $a_{m(n)}$ form a decreasing sequence bounded below by $-A$. A simple variation on the fundamental axiom (see Exercise 7.4.7 if necessary) tells us that there exists an a with $a_{m(n)} \to a$ as $n \to \infty$. Combining the results of the last two paragraphs gives the required result. ∎

Our next equivalence makes use of the definition which follows a preliminary remark.

Lemma 7.6.3 *We work in an ordered field $\mathbb{F}$. Suppose that A is a non-empty set and a_i satisfies the following properties for $i = 0, 1$.*

$(1)_i$ $a_i \ge a$ for all $a \in A$.
$(2)_i$ If $b \ge a$ for all $a \in A$, then $b \ge a_i$.

Then $a_0 = a_1$.

Proof. By $(1)_1$, $a_1 \ge a$ for all $a \in A$. Thus, by $(2)_0$, $a_1 \ge a_0$. Similarly, $a_0 \ge a_1$, so $a_0 = a_1$. ∎

Definition 7.6.4 *We work in an ordered field $\mathbb{F}$ and suppose that A is a non-empty set in $\mathbb{F}$. We say that a_0 is the* supremum *of A if a_0 satisfies the following conditions:*

(1) $a_0 \ge a$ for all $a \in A$,
(2) If $b \ge a$ for all $a \in A$, then $b \ge a_0$.

We write $\sup_{a \in A} = a_0$.

(Notice that Lemma 7.6.3 tells us that, if a supremum exists, it is unique. We can thus talk about *the* supremum.) The supremum is also called the *least upper bound* on the not unreasonable grounds that this is exactly what it is.

Exercise 7.6.5 *Let $\mathbb{F}$ be an ordered field. Consider the following choices of A and state with reasons if the supremum exists and, if it does exist, whether it lies in A.*

(i) A consists of those $a \in \mathbb{F}$ with $a \geq 1$.
(ii) A consists of those $a \in \mathbb{F}$ with $a \leq 1$.
(iii) A consists of those $a \in \mathbb{F}$ with $a < 1$.

Definition 7.6.6 *Let $\mathbb{F}$ be an ordered field. We say that $\mathbb{F}$ has the* supremum property *if, whenever A is non-empty subset of $\mathbb{F}$ which is bounded above (that is to say, there exists a K with $K \geq a$ for all $a \in A$), the set A has a supremum.*

In other words, $\mathbb{F}$ has the supremum property if any non-empty subset, which has an upper bound, has a least upper bound.

Exercise 7.6.7 *We work in an ordered field $\mathbb{F}$. We say that e_0 is the* infimum *(or* greatest lower bound*) of E if e_0 satisfies the following conditions:*

(1) $e_0 \leq e$ for all $e \in E$,
(2) If $c \leq e$ for all $e \in E$, then $c \leq e_0$.

By considering the set A of $-e$ with $e \in E$, or otherwise, show that, if $\mathbb{F}$ has the supremum property, then every non-empty set bounded below has an infimum.

Theorem 7.6.8 *Let $\mathbb{F}$ be an ordered field. The following two statements are equivalent.*

(i) $\mathbb{F}$ has the supremum property.
(ii) $\mathbb{F}$ obeys the fundamental axiom of analysis.

Proof. We split the proof into two parts.
The supremum property implies the fundamental axiom.
Suppose that $\mathbb{F}$ satisfies (i) and (a_j) is an increasing sequence bounded above. If we write A for the set consisting of the members of the sequence (a_j), then A is bounded above and so, by the supremum property, has a supremum a.

By definition,

(1) $a \geq a_j$ for all $j \geq 1$.
(2) If $b \geq a_j$ for all $j \geq 1$, then $b \geq a$.

We claim that $a_n \to a$. To see this, suppose $\epsilon > 0$. Since $a > a - \epsilon$, condition (2) tells us that there exists an N such that $a_N > a - \epsilon$. Condition (1), together with the fact that we have an increasing sequence, tells us that

$$a - \epsilon < a_N \leq a_n \leq a$$

and so $|a_n - a| < \epsilon$ for all $n \geq N$. The required result follows.
The fundamental axiom implies the supremum property.
We recall that Theorem 7.4.8 tells us that the fundamental axiom implies the axiom of Archimedes.

Let A be a non-empty set with $a < K$ for all $a \in A$. Clearly we may suppose $K > 0$. Since A is non-empty, we can find an $a_0 \in A$ and, using the axiom of Archimedes (see Exercise 7.4.6 if necessary), we can then find an integer u_0 with $u_0 \leq a_0$. For each strictly positive integer n, we consider the set E_n of integers k such that $2^{-n}k \leq a$ for some $a \in A$. The set E_n is non-empty since $u_0 2^n \in E_n$ and it is bounded above since $k < K 2^n$ whenever $k \in E_n$. Since any non-empty collection of integers which is bounded above has a largest element, E_n has a largest element which we call k_n. We set $b_n = 2^{-n}k_n$ and make the following observations.

$(1)_n$ For each n, we can find some $a_n \in A$ such that $b_n \leq a_n$.
$(2)_n$ $b_n + 2^{-n} > a$ for all $a \in A$.

We first look at inequality $(2)_n$. This tells us that $k2^{-(n+1)} < b_n + 2^{-n}$ whenever $k \in E_{n+1}$ so that $k_{n+1} \leq 2k_n + 1$ and $b_{n+1} \leq b_n + 2^{-n-1}$. A simple induction now gives $b_m \leq b_n + 2^{-n}$ for all $m \geq n$ and, in particular, $b_m \leq b_1 + 1$ for all $m \geq 1$. On the other hand, looking at the definition of k_n, we see that $2k_n \in E_{n+1}$, so $2k_n \leq k_{n+1}$ and $b_n \leq b_{n+1}$. Thus the b_n form an increasing sequence bounded above and, by the fundamental axiom, $b_n \to b$ as $n \to \infty$ for some $b \in \mathbb{F}$. We claim that b is the supremum of A.

To prove this, we first observe that, if $a \in A$, then, by the definition of b_n, we have $b_n + 2^{-n} \geq a$. Allowing $n \to \infty$ and using the axiom of Archimedes to show that $2^{-n} \to 0$, we obtain $b = b + 0 \geq a$. Thus b is indeed an upper bound.

Now suppose $c \geq a$ for all $a \in A$. By inequality $(1)_n$, we have

$$c \geq a_n \geq b_n$$

for each $n \geq 1$. Allowing $n \to \infty$, we obtain $c \geq b$. Thus b is the least upper bound, that is to say, the supremum of A. ∎

Our final equivalence involves the notion of a Cauchy sequence.

Definition 7.6.9 *We work in an ordered field* $\mathbb{F}$. *We say that a sequence* (a_n) *is a Cauchy sequence if, given* $\epsilon > 0$, *we can find an* N *such that* $|a_n - a_m| < \epsilon$ *for all* $n, m \geq N$.

The next exercise is easy, but important.

Exercise 7.6.10 *We work in an ordered field* $\mathbb{F}$. *Show that, if* $a_n \to a$ *as* $n \to \infty$, *then the* a_n *form a Cauchy sequence.*

Thus every convergent sequence is a Cauchy sequence.

Definition 7.6.11 *We say that an ordered field* $\mathbb{F}$ *is* complete (*or 'Cauchy complete'*) *if every Cauchy sequence converges, that is to say, if, whenever* (a_n) *is a Cauchy sequence, there exists an* $a \in \mathbb{F}$ *such that* $a_n \to a$ *as* $n \to \infty$.

The following remarks are very useful when dealing with Cauchy sequences.

Lemma 7.6.12 *We work in an ordered field* $\mathbb{F}$. *If some subsequence of a Cauchy sequence* (a_n) *converges to a limit* a, *then the sequence converges to the limit* a.

Proof. Suppose that we have a Cauchy sequence (a_n) and a sequence $n(1) < n(2) < \cdots$ such that $a_{n(j)} \to a$ as $j \to \infty$. Given any $\epsilon > 0$, the definition of a Cauchy sequence tells us that we can find an N such that $|a_n - a_m| < \epsilon/2$ for all $n, m > N$. On the other hand, the definition of a limit tells us that we can find a J such that $n(J) > N$ and $|a_{n(j)} - a| < \epsilon/2$ for all $j \geq J$. Thus, if $n \geq N$, we have

$$|a - a_n| \leq |a_n - a_{n(J)}| + |a_{n(J)} - a| < \epsilon/2 + \epsilon/2 = \epsilon.$$

Since ϵ was arbitrary, we are done.   ■

Exercise 7.6.13 *We work in an ordered field* $\mathbb{F}$. *Show that any Cauchy sequence* (a_n) *is bounded (that is to say, there exists a constant* K *such that* $|a_n| \leq K$ *for every* n).

Theorem 7.6.14 *Let* $\mathbb{F}$ *be an ordered field. The following two statements are equivalent.*

(i) $\mathbb{F}$ *is complete and obeys the axiom of Archimedes.*
(ii) $\mathbb{F}$ *obeys the fundamental axiom of analysis.*

Proof. We split the proof into two parts.

The fundamental axiom implies completeness and the Archimedean axiom
Suppose the fundamental axiom holds. We have already seen that the fundamental axiom implies the Archimedean axiom (see Theorem 7.4.8). We also know (from Theorem 7.6.2) that the fundamental axiom implies the Bolzano–Weierstrass condition. We shall use the Bolzano–Weierstrass condition to prove that $\mathbb{F}$ is complete as follows.

Suppose that we have a Cauchy sequence (a_n). Then the sequence must be bounded (Exercise 7.6.13) and so, by the Bolzano–Weierstrass property, must have a convergent subsequence. By Lemma 7.6.12, any Cauchy sequence with a convergent subsequence converges, so we are done.

Completeness and the Archimedean axiom[25] together imply the fundamental axiom
Let $\mathbb{F}$ be an ordered field which is complete and satisfies the Archimedean axiom. The desired result will follow if we can show any increasing sequence bounded above is a Cauchy sequence and so converges by completeness.

Suppose therefore that (a_n) is an increasing sequence bounded above by M. If it were not a Cauchy sequence we could find an $\epsilon > 0$ such that, given any N, we could find $m > n > N$ with $|a_m - a_n| > \epsilon$ and so (since we have an increasing sequence) $a_m > a_N + \epsilon$. Thus, given any J, we can find

$$1 = n(0) < n(1) < n(2) < \cdots < n(J + 1) \text{ such that } a_{n(j+1)}$$
$$-a_{n(j)} \geq \epsilon \text{ for } 0 \leq j \leq J.$$

It follows by a simple induction that

$$a_{n(J)} \geq J\epsilon + a_1,$$

so

$$M - a_1 \geq J\epsilon$$

for all J, contradicting the Archimedean axiom in the form given by Exercise 7.4.5. ■

Exercise 7.6.15 *Logically speaking, this exercise is unnecessary, since it follows from other results that $\mathbb{Q}$ does not satisfy the fundamental axiom of analysis and so cannot satisfy any of its equivalent forms. However, the reader*

---

[25] The reader may ask if we cannot deduce the Archimedean axiom from completeness. It is beyond the scope of this book, but there exist ordered fields $\mathbb{F}$ such that every Cauchy sequence (a_n) is eventually constant (that is to say, there exists an N such that $a_n = a_N$ for $n \geq N$). Since every convergent sequence is Cauchy, the convergent sequences for such a field are exactly those which are eventually constant. The field is complete (though for an uninteresting reason), but does not satisfy the axiom of Archimedes.

may find it helpful to prove these results explicitly. Recall that the equation $x^2 = 2$ has no solution in $\mathbb{Q}$ (see Theorem 4.4.11 if necessary). Throughout, we work in the ordered field $\mathbb{Q}$.

(i) *(Easy) Show that $\mathbb{Q}$ satisfies the axiom of Archimedes.*

(ii) *Explain why we can find a strictly positive integer r_n with the property that $r_n^2 \leq 2^{2n+1} < (r_n + 1)^2$.*

(iii) *Set $a_n = r_n 2^{-n}$. Show that, if $a_n \to a$ as $n \to \infty$, then $a^2 = 2$. Deduce that the sequence (a_n) has no limit.*

(iv) *Show that the sequence (a_n) is increasing and bounded above. Conclude that $\mathbb{Q}$ does not satisfy the fundamental axiom of analysis.*

(v) *Let A be the set consisting of the points a_n $[n \geq 1]$. Show that, if $b \in \mathbb{Q}$ and $b \geq a_n$ for all n, then $b^2 \geq 2$ and so $b^2 > 2$. If we set*

$$c = \frac{1}{2}\left(b + \frac{2}{b}\right)$$

(an idea that goes back to the ancient Babylonians), show that $c \in \mathbb{Q}$ with $b > c$ and $c^2 \geq 2$. Show that $c \geq a_n$ for all n and deduce that A has no supremum.

(vi) *Show that $a_n \leq a_{n+1} \leq a_n + 2^{-n}$. Deduce that $|a_n - a_m| \leq 2^{-n+1}$ for all $m \geq n$ and so the a_n form a Cauchy sequence with no limit.*

8

And What Is Its Solution?

8.1 A Construction of the Real Numbers

We have seen, in Theorem 7.4.3, that, if we wish to base calculus on limits, use the associated definition of continuous function and demand that the intermediate value condition holds, we need an ordered field which obeys the fundamental axiom of analysis.

We shall now construct such a field and show that it is unique (up to the appropriate isomorphism). We call this field the field of *real numbers*.

When we constructed the strictly positive rationals, we looked at 'equivalence classes of fractional representations'. When we constructed the rationals, we looked at 'equivalence classes of representations by differences'. Now we look at 'equivalence classes of representations as limits of sequences of rationals'.

More formally, we look at S, the collection of Cauchy sequences in $\mathbb{Q}$

$$\mathbf{a} = (a_1, a_2, a_3, \ldots).$$

(We can think of the a_j as successive rational approximations to a putative 'real number'.)

Exercise 8.1.1 *If* $\mathbf{a}$, $\mathbf{b} \in S$, *we say that* $\mathbf{a} \sim \mathbf{b}$ *if*

$$a_n - b_n \to 0 \text{ as } n \to \infty.$$

Show that $\sim$ *is an equivalence relation. (See Definition 2.2.4.)*

We shall write [$\mathbf{a}$] for the equivalence class of $\mathbf{a}$.

The definition of addition and multiplication and the proof that they give a field follow the well-worn track of our constructions of the strictly positive rationals and the rationals.

167

Lemma 8.1.2

(i) If **a**, **b** $\in \mathcal{S}$ *and we write*

$$\mathbf{a} + \mathbf{b} = (a_1 + b_1, a_2 + b_2, \ldots),$$

then **a** + **b** $\in \mathcal{S}$.

(ii) If **a**, **a**′, **b**, **b**′ $\in \mathcal{S}$ *and* **a** $\sim$ **a**′, **b** $\sim$ **b**′, *then*

$$\mathbf{a} + \mathbf{b} \sim \mathbf{a}' + \mathbf{b}'.$$

(iii) If **a**, **b** $\in \mathcal{S}$ *and we write*

$$\mathbf{a} \times \mathbf{b} = (a_1 \times b_1, a_2 \times b_2, \ldots),$$

then **a** × **b** $\in \mathcal{S}$.

(iv) If **a**, **a**′, **b**, **b**′ $\in \mathcal{S}$ *and* **a** $\sim$ **a**′, **b** $\sim$ **b**′, *then*

$$\mathbf{a} \times \mathbf{b} \sim \mathbf{a}' \times \mathbf{b}'.$$

Remark If the reader wishes to replace **a** + **b** by **a** $\oplus$ **b** and so on, she should feel free to do so, but I shall make much less use of such notation from now on.

Proof. We prove parts (iii) and (iv), leaving the slightly easier parts (i) and (ii) to the reader. Remember that we are working in $\mathbb{Q}$ so 'Cauchy' means 'Cauchy in $\mathbb{Q}$'.

(iii) We know that Cauchy sequences are bounded (see Exercise 7.6.13), so we can find an $M > 0$ such that $|a_j|$, $|b_j| \leq M$ for all j. If $\epsilon > 0$, then, since we have two Cauchy sequences, we can find an N such that

$$|a_m - a_n|, \; |b_m - b_n| < \frac{\epsilon}{2M}$$

for all m, $n \geq N$. It follows that

$$
\begin{aligned}
\left|(a_m \times b_m) - (a_n \times b_n)\right| &= \left|\left(a_m \times (b_m - b_n)\right) + \left(b_n \times (a_m - a_n)\right)\right| \\
&\leq \left|a_m \times (b_m - b_n)\right| + \left|b_n \times (a_m - a_n)\right| \\
&= \left(|a_m| \times |b_m - b_n|\right) + \left(|b_n| \times |a_m - a_n|\right) \\
&< M \frac{\epsilon}{2M} + M \frac{\epsilon}{2M} = \epsilon
\end{aligned}
$$

for all m, $n \geq N$. Thus **a** × **b** is a Cauchy sequence.

(iv) We use a similar argument to (iii). We know that Cauchy sequences are bounded, so we can find an $M > 0$ such that $|a_j|$, $|a_j'|$, $|b_j|$, $|b_j'| \leq M$ for all j. If $\epsilon > 0$, then, since **a** $\sim$ **a**′ and **b** $\sim$ **b**′, we can find an N such that

$$|a_n - a_n'|, \; |b_n - b_n'| < \frac{\epsilon}{2M}$$

for all $n \geq N$. It follows that

$$
\begin{aligned}
\left| (a_n \times b_n) - (a'_n \times b'_n) \right| &= \left| \left(a_n \times (b_n - b'_n) \right) + \left(b'_n \times (a_n - a'_n) \right) \right| \\
&\leq \left| a_n \times (b_n - b'_n) \right| + \left| b'_n \times (a_n - a'_n) \right| \\
&= \left(|a_n| \times |b_n - b'_n| \right) + \left(|b'_n| \times |a_n - a'_n| \right) \\
&< M \frac{\epsilon}{2M} + M \frac{\epsilon}{2M} = \epsilon
\end{aligned}
$$

for all $n \geq N$. Since ϵ was arbitrary, we have shown that

$$\mathbf{a} \times \mathbf{b} \sim \mathbf{a}' \times \mathbf{b}',$$

as required. ∎

Exercise 8.1.3 *Prove parts (i) and (ii) of Lemma 8.1.2.*

We can thus make the following definition.

Definition 8.1.4 *Using the notation of Lemma 8.1.2, we write*

$$[\mathbf{a}] + [\mathbf{b}] = [\mathbf{a} + \mathbf{b}], \ [\mathbf{a}] \times [\mathbf{b}] = [\mathbf{a} \times \mathbf{b}].$$

Let $\mathbf{0} = (0, 0, 0, \ldots)$ and $\mathbf{1} = (1, 1, 1, \ldots)$. We observe that $\mathbf{0}, \mathbf{1} \in S$.

Lemma 8.1.5 *If we write $\mathbb{R}$ for the collection of equivalence classes $[\mathbf{a}]$, then $\mathbb{R}$ with addition $+$, multiplication $\times$, additive zero $[\mathbf{0}]$ and multiplicative unit $[\mathbf{1}]$ is a field (see Definition 5.1.1).*

Proof. All the verifications are easy apart from condition (viii), that is to say, the existence of a multiplicative inverse. We shall prove this condition in a separate lemma below.

For the moment, we prove two of the other conditions, leaving it to the reader to check the rest.

(iv) (Additive inverse) Let $\mathbf{a} \in S$. If we write

$$-\mathbf{a} = (-a_1, -a_2, -a_3, \ldots),$$

then, since $|(-a_n) - (-a_m)| = |a_n - a_m|$, we have $-\mathbf{a} \in S$. Automatically

$$[\mathbf{a}] + [-\mathbf{a}] = [\mathbf{a} + (-\mathbf{a})] = [\mathbf{0}].$$

(ix) (Distributive law) The distributive law for $\mathbb{Q}$ gives

$$a_n \times (b_n + c_n) = (a_n \times b_n) + (a_n \times c_n)$$

for all n, so $\mathbf{a} \times (\mathbf{b} + \mathbf{c}) = (\mathbf{a} \times \mathbf{b}) + (\mathbf{a} \times \mathbf{c})$ and

$$[\mathbf{a}] \times ([\mathbf{b}] + [\mathbf{c}]) = ([\mathbf{a}] \times [\mathbf{b}]) + ([\mathbf{a}] \times [\mathbf{c}]). \ ∎$$

Exercise 8.1.6 *Verify the remaining conditions [apart from (viii)].*

The proof of condition (viii) in Lemma 8.1.5 follows an expected path, but we have to be careful since we must avoid division by zero and we have to ensure that division by small numbers does not produce unexpected large numbers.

Lemma 8.1.7 *We use the hypotheses and notation of Lemma 8.1.5.*

(i) *If* [**a**] $\neq$ [**0**], *we can find strictly positive integers M and N such that* $|a_n| \geq 1/M$ *for all* $n \geq N$.

(ii) *If* [**a**] $\neq$ [**0**], *we can find an* **a**$'$ $\in \mathcal{S}$ *and a strictly positive integer M such that* **a**$'$ $\sim$ **a** *and* $|a_n'| \geq 1/M$ *for all n.*

(iii) *Suppose that* [**a**] $\neq$ [**0**] *and* **a**$'$ *satisfies the conclusions of (ii). Then, if we write $b_n = 1/a_n'$, we have* **b** $\in \mathcal{S}$ *and*

$$[\mathbf{a}] \times [\mathbf{b}] = [\mathbf{1}].$$

Proof. (i) Suppose that **a** $\in \mathcal{S}$, but there do not exist strictly positive integers M and N such that $|a_n| \geq 1/M$ for all $n \geq N$. We can then find $n(1) < n(2) < n(3) < \cdots$ such that

$$|a_{n(j)}| < 1/j.$$

By the axiom of Archimedes for $\mathbb{Q}$, it follows that $a_{n(j)} \to 0$ as $j \to \infty$. We know that, if any subsequence of a Cauchy sequence converges to a limit, then the sequence converges to that limit (see Lemma 7.6.12), so $a_n \to 0$ as $n \to \infty$. Thus **a** $\sim$ **0** and [**a**] = [**0**].

(ii) Let M and N be as in (i). Set

$$a_n' = \begin{cases} 1/M & \text{if } 1 \leq n \leq N, \\ a_n & \text{otherwise.} \end{cases}$$

(iii) Let $\epsilon > 0$. Since **a**$'$ is a Cauchy sequence we can find a strictly positive integer P such that

$$|a_n' - a_m'| < \frac{\epsilon}{M^2}$$

for all $n, m \geq P$. Automatically

$$\left| \frac{1}{a_n'} - \frac{1}{a_m'} \right| = \left| \frac{a_m' - a_n'}{a_n' \times a_m'} \right| = \frac{|a_m' - a_n'|}{|a_n'| \times |a_m'|} < \frac{\epsilon}{M^2} \times M^2 = \epsilon$$

for all $n, m \geq P$. We have shown that **b** $\in \mathcal{S}$.

We finish the proof by observing that

$$[\mathbf{a}] \times [\mathbf{b}] = [\mathbf{a}'] \times [\mathbf{b}] = [\mathbf{a}' \times \mathbf{b}] = [\mathbf{1}]. \qquad \blacksquare$$

We now seek to define an order on $\mathbb{R}$. The reader will see that our sequence of temporary definition, lemma and final definition follows a standard pattern.

Definition 8.1.8 *Let* $\mathbf{a}$, $\mathbf{b} \in S$. *We say that* $\mathbf{a} \succ \mathbf{b}$ *if there exist strictly positive integers M and N such that*

$$a_j \geq b_j + \frac{1}{M}$$

for all $j \geq N$.

Lemma 8.1.9 *Let* $\mathbf{a}$, $\mathbf{a}'$, $\mathbf{b}$, $\mathbf{b}' \in S$. *If* $\mathbf{a} \sim \mathbf{a}'$, $\mathbf{b} \sim \mathbf{b}'$ *and* $\mathbf{a} \succ \mathbf{b}$, *then* $\mathbf{a}' \succ \mathbf{b}'$.

Proof. By definition, there exist strictly positive integers M and N such that

$$a_j \geq b_j + \frac{1}{M}$$

for all $j \geq N$. Since $\mathbf{a} \sim \mathbf{a}'$ and $\mathbf{b} \sim \mathbf{b}'$, we can find an $N' \geq N$ such that

$$|a_j - a'_j|, \ |b_j - b'_j| \leq \frac{1}{3M}$$

for all $j \geq N'$.

We now have

$$a'_j \geq a_j - \frac{1}{3M} \geq b_j + \frac{2}{3M} \geq b'_j + \frac{1}{3M}$$

for all $j \geq N'$ and so $\mathbf{a}' \succ \mathbf{b}'$. $\qquad \blacksquare$

Lemma 8.1.9 allows us to make the following definition.

Definition 8.1.10 *Let* $\mathbf{a}$, $\mathbf{b} \in S$. *We say that* $[\mathbf{a}] > [\mathbf{b}]$ *if* $\mathbf{a} \succ \mathbf{b}$.

Lemma 8.1.11 *The field* $(\mathbb{R}, +, \times)$ *obtained in this section is an ordered field for the inequality $>$ of Definition 8.1.10. More specifically,*

(x) *If* $[\mathbf{a}] > [\mathbf{b}]$ *and* $[\mathbf{b}] > [\mathbf{c}]$, *then* $[\mathbf{a}] > [\mathbf{c}]$. *(Transitivity of order)*

(xi) *Exactly one of the following conditions holds:* $[\mathbf{a}] > [\mathbf{b}]$ *or* $[\mathbf{b}] > [\mathbf{a}]$ *or* $[\mathbf{a}] = [\mathbf{b}]$. *(Trichotomy)*

(xii) *If* $[\mathbf{a}] > [\mathbf{b}]$, *then* $[\mathbf{a}] + [\mathbf{c}] > [\mathbf{b}] + [\mathbf{c}]$. *(Order and addition)*

(xiii) *If* $[\mathbf{a}] > [\mathbf{b}]$ *and* $[\mathbf{c}] > [\mathbf{0}]$, *then* $[\mathbf{a}] \times [\mathbf{c}] > [\mathbf{b}] \times [\mathbf{c}]$. *(Order and multiplication)*

Proof. We leave the proof of (x), (xii) and (xiii) to the reader and concentrate on proving (xi).

We need to show that at most one of the relations [**a**] > [**b**] or [**b**] > [**a**] or [**a**] = [**b**] can hold. First we show that, if [**a**] > [**b**], then it is not the case that [**a**] = [**b**]. For, if [**a**] > [**b**], then there exist strictly positive integers M and N such that

$$a_j \geq b_j + \frac{1}{M}$$

and so $|a_j - b_j| \geq 1/M$ for all $j \geq N$. In particular, $a_j - b_j \nrightarrow 0$ as $j \rightarrow \infty$, so

$$[\mathbf{a}] - [\mathbf{b}] = [\mathbf{a} - \mathbf{b}] \neq [\mathbf{0}]$$

and [**a**] $\neq$ [**b**]. Similar arguments show that if [**a**] > [**b**], then it is not the case that [**b**] > [**a**]. The symmetry between [**b**] and [**a**] covers the remaining cases.

We now need to show that at least one of the relations [**a**] > [**b**] or [**b**] > [**a**] or [**a**] = [**b**] must hold. Suppose, therefore, that neither of the conditions [**a**] > [**b**] and [**b**] > [**a**] hold.

If $\epsilon > 0$, then we can find a strictly positive integer M such that $\epsilon/3 \geq 1/M$. Since **a** and **b** are Cauchy sequences, we can find an N such that

$$|a_n - a_m|, \ |b_n - b_m| < \frac{1}{M}$$

for all n, $m \geq N$. Since it is not the case that [**a**] > [**b**], we can find an $n' \geq N$ such that

$$a_{n'} < b_{n'} + \frac{1}{M}.$$

We thus have

$$a_n < |a_n - a_{n'}| + a_{n'} < \frac{2}{M} + b_{n'} < \frac{2}{M} + b_n + |b_n - b_{n'}| < \frac{3}{M} + b_n$$

for all $n \geq N$. Similarly, since it is not the case that [**a**] < [**b**], we have

$$b_n < \frac{3}{M} + a_n$$

for all $n \geq N$. We have shown that

$$|a_n - b_n| < \frac{3}{M} < \epsilon$$

for all $n \geq N$. Since ϵ was chosen freely, we have $a_n - b_n \rightarrow 0$ as $n \rightarrow \infty$, so

$$[\mathbf{a}] - [\mathbf{b}] = [\mathbf{a} - \mathbf{b}] = [\mathbf{0}],$$

that is to say, [**a**] = [**b**]. ∎

Exercise 8.1.12 *Prove conditions* (x), (xii) *and* $(xiii)$ *of Lemma 8.1.11.*

So far, we have dealt with the algebraic aspects of $(\mathbb{R}, +, \times, >)$, showing that it is an ordered field. We now turn to the 'analysis' side of the argument and seek to show that our ordered field satisfies the fundamental axiom of analysis.

Before beginning the argument proper, we have to resolve a notational difficulty. We saw earlier that every ordered field contains a unique copy of $\mathbb{Q}$, so it is natural to call this copy $\mathbb{Q}$. However, when we seek to construct $\mathbb{R}$ from $\mathbb{Q}$ we need to distinguish the copy from the original. To do this, we write

$$\mathbf{u}(q) = (q, q, q, \ldots)$$

where $q \in \mathbb{Q}$, that is to say, we write $\mathbf{u}(q)$ for the sequence each element of which is q.

Exercise 8.1.13 *Check that the following statements are true. We take* q, $q' \in \mathbb{Q}$.

(i) *We have* $\mathbf{u}(q) \in \mathcal{S}$.

(ii) *The function* $f : \mathbb{Q} \to \mathbb{R}$ *given by* $f(q) = [\mathbf{u}(q)]$ *is injective.*

(iii) *We have* $f(q + q') = f(q) + f(q')$, $f(q \times q') = f(q) \times f(q')$ *and, whenever* $q > q'$, *we have* $f(q) > f(q')$.

Thus the $[\mathbf{u}(q)]$ give the model of $\mathbb{Q}$ in $\mathbb{R}$. Note also that $[\mathbf{u}(0)] = [\mathbf{0}]$ and $[\mathbf{u}(1)] = [\mathbf{1}]$.

Lemma 8.1.14 [**The Archimedean property for** $\mathbb{R}$] *We have* $[\mathbf{u}(1/n)] \to [\mathbf{0}]$ *as* $n \to \infty$.

Proof. Suppose that $[\epsilon] > [\mathbf{0}]$. Then, by definition, we can find strictly positive integers M and N such that

$$\epsilon_j \geq \frac{1}{M}$$

for all $j \geq N$. Thus

$$\epsilon_j > \frac{1}{2M} + \frac{1}{n}$$

for all $j \geq N$ and $n \geq 2M + 1$. We have shown that

$$[\epsilon] > [\mathbf{u}(1/n)] > [\mathbf{u}(0)] = [\mathbf{0}]$$

for $n \geq 2M + 1$. Since $[\epsilon]$ was arbitrary, it follows that $[\mathbf{u}(1/n)] \to [\mathbf{0}]$ as $n \to \infty$. ∎

Theorem 7.4.11 has the following useful consequence.

Lemma 8.1.15 *If* $[\mathbf{a}] \in \mathbb{R}$ *and* $[\epsilon] > [\mathbf{0}]$, *then we can find some* $q \in \mathbb{Q}$ *such that*

$$|[\mathbf{a}] - [\mathbf{u}(q)]| < [\epsilon].$$

We also need the following observation.

Lemma 8.1.16 *If* $\mathbf{q} = (q_1, q_2, q_3, \ldots) \in \mathcal{S}$, *then* $[\mathbf{u}(q_j)] \to [\mathbf{q}]$ *as* $j \to \infty$.

Proof. Suppose $[\epsilon] > [\mathbf{0}]$. By the axiom of Archimedes in the form given in Exercise 7.4.5, we can find a strictly positive integer M such that $[\epsilon] > [\mathbf{u}(1/M)]$. Since $\mathbf{q} \in \mathcal{S}$, we can find an N such that $|q_n - q_m| < 1/M$ for all $n, m \geq N$. We thus have

$$|u_m(q_n) - q_m| = |q_n - q_m| < 1/M$$

for all $n, m \geq N$ and so

$$|[\mathbf{u}(q_n)] - [\mathbf{q}]| = |[\mathbf{u}(q_n)] - \mathbf{q}| \leq [\mathbf{u}(1/M)] < [\epsilon]$$

for all $n \geq N$. The required result follows. ∎

Theorem 8.1.17 *The ordered field* $(\mathbb{R}, +, \times, >)$ *satisfies the fundamental axiom of analysis.*

Proof. We already know that $\mathbb{R}$ satisfies the axiom of Archimedes, so, by Theorem 7.6.14, we need only show that $\mathbb{R}$ is complete.

To this end, suppose that $[\mathbf{a}(1)], [\mathbf{a}(2)], [\mathbf{a}(3)], \ldots$ form a Cauchy sequence in $\mathbb{R}$. (Thus we have a Cauchy sequence in $\mathbb{R}$ of equivalence classes of Cauchy sequences in $\mathbb{Q}$.) By Lemma 8.1.15, we can find $q_n \in \mathbb{Q}$ such that

$$|[\mathbf{a}(n)] - [\mathbf{u}(q_n)]| < \mathbf{u}(1/n)$$

for each n. We claim that the q_n form a Cauchy sequence in $\mathbb{Q}$.

Suppose $\epsilon \in \mathbb{Q}$ and $\epsilon > 0$. We can find a strictly positive integer M such that $\epsilon/3 > 1/M$. Since $[\mathbf{a}(n)]$ is a Cauchy sequence in $\mathbb{R}$, we can find an $N > M$ such that

$$|[\mathbf{a}(n)] - [\mathbf{a}(m)]| < \mathbf{u}(1/M)$$

for all $n, m \geq N$. We now have

$$
\begin{aligned}
|[\mathbf{u}(q_n - q_m)]| &= |[\mathbf{u}(q_n)] - [\mathbf{u}(q_m)]| \\
&\leq |[\mathbf{u}(q_n)] - [\mathbf{a}(m)]| + |[\mathbf{u}(q_m)] - [\mathbf{a}(m)]| + |[\mathbf{a}(n)] - [\mathbf{a}(m)]| \\
&\leq \mathbf{u}(1/n) + \mathbf{u}(1/m) + \mathbf{u}(1/M) < \mathbf{u}(3/M)
\end{aligned}
$$

and so $|q_n - q_m| \leq 3/M < \epsilon$ for all n, $m \geq N$. We have proved that the q_n form a Cauchy sequence in $\mathbb{Q}$ and so, in particular,

$$\mathbf{a} = (q_1, q_2, q_3, \ldots) \in \mathcal{S}.$$

By Lemma 8.1.16,

$$\mathbf{u}(q_n) \to \mathbf{a}$$

and, by the choice of q_n made in the second paragraph of the proof (together with the axiom of Archimedes),

$$[\mathbf{a}(n)] - [\mathbf{u}(q_n)] \to [\mathbf{0}]$$

as $n \to \infty$. Thus

$$[\mathbf{a}(n)] = [\mathbf{u}(q_n)] + \big([\mathbf{a}(n)] - [\mathbf{u}(q_n)]\big) \to [\mathbf{a}] + [\mathbf{0}] = [\mathbf{a}]$$

and we have proved that $\mathbb{R}$ is complete. ∎

We have shown how to construct an ordered field satisfying the fundamental axiom, that is to say, an object on which we can do calculus in the Cauchy manner. We now show that (with our standard disclaimer 'up to isomorphism') there is only one such field. We start with some simple exercises.

Exercise 8.1.18 *We shall take* $(\mathbb{F}, +, \times, >)$ *and* $(\mathbb{G}, +, \times, >)$ *to be ordered fields. We suppose that* $h : \mathbb{F} \to \mathbb{G}$ *is a bijection which preserves* $+$, $\times$ *and* $>$.

(i) *Show that* $h^{-1} : \mathbb{G} \to \mathbb{F}$ *preserves* $+$, $\times$ *and* $>$. *[See the proof of part (i) of Lemma 3.4.14 if you need a hint.]*
(ii) *Show that, if* $a \in \mathbb{F}$, *then* $|h(a)| = h(|a|)$.
(iii) *Show that, if* $a_n \to a$ *as* $n \to \infty$ *in* $\mathbb{F}$, *then* $h(a_n) \to h(a)$ *in* $\mathbb{G}$.
(iv) *If the sequence* (x_n) *is Cauchy in* $\mathbb{F}$, *show that the sequence* $\big(h(x_n)\big)$ *is Cauchy in* $\mathbb{G}$.

Theorem 8.1.19 *Suppose that* $(\mathbb{F}, +, \times, >)$ *and* $(\mathbb{G}, +, \times, >)$ *are ordered fields satisfying the fundamental axiom. Then there exists a bijection* $h : \mathbb{F} \to \mathbb{G}$ *which preserves* $+$, $\times$ *and* $>$.

Proof. Exercise 7.2.7 tells us that $\mathbb{F}$ and $\mathbb{G}$ contain a 'natural copy' of $\mathbb{Q}$. More precisely, there exist injective functions, $f : \mathbb{Q} \to \mathbb{F}$ and $g : \mathbb{Q} \to \mathbb{G}$ which preserve $+$, $\times$ and $>$. Let us write $\mathbb{Q}_\mathbb{F}$ for the image of $\mathbb{Q}$ under f (that is to say, the collection of $f(q)$ with $q \in \mathbb{Q}$) and $\mathbb{Q}_\mathbb{G}$ for the image of $\mathbb{Q}$ under g.

By Theorem 7.4.11, every element of $\mathbb{F}$ is the limit of a sequence in $\mathbb{Q}_\mathbb{F}$. Suppose that $f(q_n) \to u$ for some sequence $q_n \in \mathbb{Q}$. Then $f(q_n)$ is Cauchy in $\mathbb{F}$ and so in $\mathbb{Q}_\mathbb{F}$. It follows from Exercise 8.1.18 (iv) that the q_n form a Cauchy sequence in $\mathbb{Q}$, so the $g(q_n)$ form a Cauchy sequence in $\mathbb{Q}_\mathbb{G}$ and so (since we have the axiom of Archimedes) in $\mathbb{G}$. Thus $g(q_n)$ converges to some limit v in $\mathbb{G}$.

Now suppose $f(q'_n) \to u$ for some sequence of $q'_n \in \mathbb{Q}$. We have

$$|f(q_n - q'_n)| = |f(q_n) - f(q'_n)| \le |f(q_n) - u| + |f(q'_n) - u| \to 0$$

as $n \to \infty$. It follows from Exercise 8.1.18 (iii) that $q_n - q'_n \to 0$ as $n \to \infty$ and so $g(q_n) - g(q'_n) \to 0$. Thus

$$g(q'_n) = u + \big(g(q'_n) - g(q_n)\big) + \big(g(q_n) - u\big) \to u + 0 + 0 = u$$

as $n \to \infty$.

We can thus define a unique function $h : \mathbb{F} \to \mathbb{G}$ by the condition that, if $f(q_n) \to u$ for some sequence of $q_n \in \mathbb{Q}$, then $h(u)$ is the limit of $g(q_n)$. Similarly, there exists a unique function $k : \mathbb{G} \to \mathbb{F}$ given by the condition that, if $g(q_n) \to v$ for some sequence of $q_n \in \mathbb{Q}$, then $k(v)$ is the limit of $f(q_n)$.

We now show that h and k are bijective with $k = h^{-1}$. If $u \in \mathbb{F}$, choose a sequence of $q_n \in \mathbb{Q}$ such that $f(q_n) \to u$. By definition, $g(q_n) \to h(u)$, so, again by definition, $k\big(h(u)\big) = u$. Similarly $h\big(k(v)\big) = v$ for all $v \in \mathbb{G}$. Thus h and k are inverse functions and we are done.

Finally, we need to check that h preserves $+$, $\times$ and $>$. To see that h preserves $+$, observe that, if u, $u' \in \mathbb{F}$, then we can find sequences q_n, $q'_n \in \mathbb{Q}$ such that $f(q_n) \to u$ and $f(q'_n) \to u'$. We have $g(q_n) \to h(u)$ and $g(q'_n) \to h(u')$. On the other hand, $q_n + q'_n \in \mathbb{Q}$ and f preserves $+$, so

$$f(q_n + q'_n) = f(q_n) + f(q'_n) \to u + u';$$

so $g(q_n + q'_n) \to h(u + u')$. Since g preserves $+$,

$$g(q_n + q'_n) = g(q_n) + g(q'_n) \to h(u) + h(u')$$

and $h(u + u') = h(u) + h(u')$. A similar argument shows that $h(u \times u') = h(u) \times h(u')$.

The proof that h preserves $>$ requires a little more care.[1] Suppose that u, $u' \in \mathbb{F}$ and $u > u'$. By the axiom of Archimedes, we can find a strictly positive integer M such that $f(1/3) \times (u - u') > f(1/M)$. We now choose q_n, $q'_n \in \mathbb{Q}$ such that

$$|f(q_n) - u| \le f(1/M) \text{ and } |f(q'_n) - u'| \le f(1/M) \text{ for all } n$$

---

[1] Recall that we may have $x_n \to x$, $y_n \to y$, $x_n > y_n$ for all n, but $x = y$.

and $f(q_n) \to u$, $f(q_n') \to u'$ as $n \to \infty$. With this choice,

$$f(q_n - q_n') = f(q_n) - f(q_n') \geq \left(u - f(1/M)\right) - \left(u' + f(1/M)\right)$$
$$= (u - u') - f(2/M) \geq f(3/M) - f(2/M) = f(1/M),$$

so

$$g(q_n - q_n') \geq g(1/M).$$

Allowing $n \to \infty$, we have

$$h(u) - h(u') = h(u - u') \geq g(1/M)$$

and so $h(u) > h(u')$, as required. ∎

We call the elements of $\mathbb{R}$ the real numbers.

The construction of the real numbers from the rationals by Dedekind (and, more or less simultaneously, by others[2]) was an important achievement, but, viewed in retrospect, the main outcome of the Cauchy programme was the identification of how the real numbers and the calculus based on them actually worked. Most university courses now begin with the statement that $\mathbb{R}$ is an ordered field satisfying the fundamental axiom and proceed from there.

Exercise 8.1.20 *This easy exercise completes unfinished business from Exercise 7.2.10. We know (by Lemma 7.4.12) that $\mathbb{R}$ contains $\sqrt{2}$. If $\mathbb{Q}[\sqrt{2}]$ is defined as in Exercise 7.2.10, show that the function $g : \mathbb{Q}[\sqrt{2}] \to \mathbb{R}$ given by $g(\mathbf{a}) = a_1 + a_2\sqrt{2}$ is an injection preserving addition and multiplication, in other words a field isomorphism of $\mathbb{Q}[\sqrt{2}]$ with*

$$\mathbb{G} = \{a_1 + a_2\sqrt{2} : a_1, a_2 \in \mathbb{Q}\}.$$

Show that $\mathbf{a} \ominus \mathbf{b}$ if and only if $g(\mathbf{a}) > g(\mathbf{b})$. Conclude that $\mathbb{Q}[\sqrt{2}]$ is an ordered field.

8.2 Some Consequences

Any university mathematics course with the words analysis in its title and many without (differential geometry, advanced probability, ...) may be viewed as the study of $\mathbb{R}$ and objects constructed from it. This section is a pause to admire the view at the beginning of a long climb.

We start by giving a 'Cauchy calculus' treatment of Example 7.3.1.

---

[2] The construction given in this section is due to Cantor. I have chosen this method because it is echoed by similar constructions in advanced analysis.

Example 8.2.1

(i) $1 + \dfrac{1}{2} + \dfrac{1}{2^2} + \dfrac{1}{2^3} + \cdots + \dfrac{1}{2^n} \to 2$ *as* $n \to \infty$.

(ii) $\dfrac{9}{10} + \dfrac{9}{10^2} + \dfrac{9}{10^3} + \cdots + \dfrac{9}{10^n} \to 1$ *as* $n \to \infty$.

(iii) $1 + 2 + 2^2 + 2^3 + \cdots + 2^n$ *does not converge as* $n \to \infty$.

Proof. We give proofs echoing Example 7.3.1.

(i) Let
$$S_n = 1 + \frac{1}{2} + \frac{1}{2^2} + \frac{1}{2^3} + \cdots + \frac{1}{2^n}.$$

Then (S_n) is an increasing sequence bounded above by 2 (since a simple induction gives $S_n = 2 - 1/2^n$). It follows that S_n converges to some limit S say. We now observe that

$$2S_{n-1} - S_n = 2;$$

so, allowing $n \to \infty$, we have $2S - S = 2$, that is to say, $S = 2$.

(ii) Left to the reader.

(iii) Let
$$S_n = 1 + 2 + 4 + 8 + \ldots + 2^n.$$

Then $|S_n - S_{n+1}| \geq 1$ so (S_n) is not Cauchy and so does not converge. ∎

Exercise 8.2.2 *Here is a more workaday method of attacking Example 8.2.1.*

(i) (Summing a geometric series) Consider a field $\mathbb{F}$ *and some* $x \in \mathbb{F}$ *with* $x \neq 1$. *If we set* $S_0(x) = 1$ *and* $S_n(x) = S_{n-1}(x) + x^n$, *show, by induction, that*

$$S_n(x) = \frac{1 - x^{n+1}}{1 - x}$$

or, more informally, that

$$1 + x + \ldots + x^n = \frac{1 - x^{n+1}}{1 - x}.$$

(ii) We now take $\mathbb{F} = \mathbb{R}$. *Use (i) and Exercise 7.4.9 to show that, if* $|x| < 1$, *then*

$$1 + x + \cdots + x^n \to \frac{1}{1 - x}$$

as $n \to \infty$ *and hence obtain the results of the first two parts of Example 8.2.1.*

We can also show that every decimal corresponds to a real number and every real number can be written in decimal form.

Exercise 8.2.3 *If a_j is an integer with $9 \geq a_j \geq 0$ for all $j \geq 1$, show that*

$$\frac{a_1}{10} + \frac{a_2}{10^2} + \cdots \frac{a_n}{10^n} \to x$$

as $n \to \infty$, for some $x \in \mathbb{R}$ with $0 \leq x \leq 1$.

Exercise 8.2.4 *If x is a real number with $1 > x \geq 0$, let us write Tx for the integer part of $10x$ (so that $Tx = m$, where m is an integer and $m + 1 > 10x \geq m$) and $Sx = 10x - Tx$. If $1 > a \geq 0$, let us set $a_1 = Ta$, $b_1 = Sa$ and then, proceeding inductively, set $a_{n+1} = Tb_n$ and $b_{n+1} = Sb_n$.*

(i) *Show that $9 \geq a_n \geq 0$ and $1 > b_n \geq 0$.*
(ii) *Let us define $x_1 = a_1 10^{-1}$ and $x_{n+1} = a_{n+1} 10^{-n-1} + x_n$. Show that $10^n a = 10^n x_n + b_n$ Conclude that $10^{-n} > a - x_n \geq 0$ and deduce, using the axiom of Archimedes, that $x_n \to a$ as $n \to \infty$.*
(iii) *Explain, in simple terms, the relevance of this construction to decimal expansion.*

Exercise 8.2.5 *We continue with the ideas and notation of Exercise 8.2.4*

(i) *Suppose that $a = u/v$ with u and v non-negative integers and $v > u \geq 0$. Show that $b_n = u_n/v$ where u_n is an integer with $v > u_n \geq 0$. By looking at possible values for u_r when $v+1 \geq r \geq 1$, show that there exist integers p and q with*

$$v + 1 \geq p > q \geq 1 \text{ and } u_p = u_q.$$

Show that $a_{m+(p-q)} = a_m$ for all $m \geq q + 1$.
(ii) *Suppose, conversely, that there exist integers t, $s \geq 1$ such that $a_{m+s} = a_m$ for all $m \geq t$. Show that a is rational.*

Although it is possible to construct the reals using decimal expansions rather than Cauchy sequences,[3] there are technical problems hinted at by the next exercise.

Exercise 8.2.6

(i) *Suppose that a, b, c, d are real numbers with $1/2 > c > 0$, $1/2 > d > 0$ and $a = c + d$, $b = c \times d$. Let a, b, c, d have decimal expansions*

$$a = .a_1 a_2 a_3 \ldots , \quad b = .b_1 b_2 b_3 \ldots , \quad c = .c_1 c_2 c_3 \ldots , \quad d = .d_1 d_2 d_3 \ldots .$$

(ii) *Does there exist an N (independent of the particular choice of c and d) such that, knowing c_j and d_j for $1 \leq j \leq N$, we can calculate a_1? Why?*

---

[3] If you feel that this is more *natural*, it is worth asking why the fact that we have eight fingers and two thumbs should enter into the construction of $\mathbb{R}$.

(iii) *Does there exist an M (independent of the particular choice of c and d) such that, knowing c_j and d_j for $1 \le j \le M$, we can calculate b_1? Why?*

So far, so routine. We now come to an extraordinary discovery of Cantor. (If you do not consider it extraordinary, you must explain why nobody before Cantor asked the appropriate question, let alone answered it.) The real numbers cannot be enumerated!

Theorem 8.2.7 *Consider any sequence of $x_n \in \mathbb{R}$. There exists a $y \in \mathbb{R}$ with $y \ne x_n$ for all $n \ge 1$,*

Proof. (This proof is closer to Cantor's original proof than the one usually given.) We define two sequences $\delta_j > 0$ and $y_j \in \mathbb{R}$ inductively. We set $\delta_0 = 1$ and $y_0 = 0$. Once δ_j and y_j have been defined, we look at x_{j+1}.

There are two possibilities. Either $y_j \ge x_{j+1}$ and we take $\delta_{j+1} = \delta_j/2$ and set $y_{j+1} = y_j + \delta_{j+1}$, or $y_j < x_{j+1}$, in which case we take

$$\delta_{j+1} = \min\{\delta_j, x_{j+1} - y_j\}/2$$

and set $y_{j+1} = y_j$.

We observe that, in both cases, $0 < \delta_{j+1} \le \delta_j/2$ and $y_j \le y_{j+1} \le y_j + \delta_{j+1}$. Using an inductive argument, we obtain $0 < \delta_{j+k} \le 2^{-k}\delta_j$ for all $k \ge 0$ and $j \ge 0$. A second inductive argument now shows that

$$y_j \le y_{j+k} \le y_j + (1 - 2^{-k})\delta_j, \qquad \bigstar$$

for all $k \ge 0$ and $j \ge 0$. Taking $j = 0$, we see that y_j form an increasing sequence bounded above by $y_0 + \delta_0 = 1$ and so converge to some limit y. By $\bigstar$, we have

$$y_j \le y \le y_j + \delta_j.$$

As in the second paragraph of our proof, there are two possibilities. Either $y_j \ge x_{j+1}$, so that

$$y \ge y_{j+1} = y_j + \delta_{j+1} \ge x_{j+1} + \delta_{j+1} > x_{j+1}$$

or $x_{j+1} > y_j$, so that

$$y \le y_{j+1} + \delta_{j+1} = y_j + \delta_{j+1} \le y_j + (x_{j+1} - y_j)/2 = (y_j + x_{j+1})/2 < x_{j+1}.$$

Thus $y \ne x_{j+1}$ for all $j \ge 0$ and we are done ∎

Cantor's theorem says that we cannot count off the real numbers in the same way as we count off the natural numbers: one, two, three, ... Mathematicians say that 'the reals are uncountable'[4] and murmur 'Toto, I've a feeling we're not in Kansas anymore.'

Note that the rational numbers $\mathbb{Q}$ can be enumerated.

Theorem 8.2.8 *There exists a sequence of $x_n \in \mathbb{Q}$ such that, if $y \in \mathbb{Q}$, there exists an n with $y = x_n$.*

Proof. Suppose that $n = 2^k 3^u 5^v$ with $k = 1$ or $k = 2$ and u, v natural numbers. (The uniqueness of prime factorisation tells us that k, u and v are uniquely defined.) We set

$$x_n = \begin{cases} \frac{u}{v} & \text{if } k = 1, \\ \frac{-u}{v} & \text{if } k = 2. \end{cases}$$

If n does not factorise as shown in the first sentence, we just set $x_n = 0$. ∎

Exercise 8.2.9 *Theorem 8.2.8 shows that there is a surjection $g : \mathbb{N} \to \mathbb{Q}$. Use this result to show that there is, in fact, a bijection $f : \mathbb{N} \to \mathbb{Q}$.*

How was it possible to start from the enumerable system of the rationals and end up with the non-enumerable system of the reals? A little thought shows that the break occurred when we considered $\mathcal{S}$, the set of all Cauchy sequences of rationals. Earlier, we mentioned that there exist a few mathematicians who consider the use of the infinite set $\mathbb{N}^+$ as illegitimate. They are now joined by another small, but respectable, group of mathematicians who are willing to talk about $\mathbb{N}^+$, but view the use of objects like 'all sequences of rationals' as a step too far.

The great majority of mathematicians are happy to talk about 'the set of all continuous functions $F : \mathbb{R} \to \mathbb{R}$', 'the set of all sequences of rationals' and similar objects. However, as I remarked earlier, experience has shown that unrestricted use of such definitions can lead to paradoxes. Here is another one.

Example 8.2.10 *Berry's paradox* *Consider the smallest positive integer not describable in an English sentence containing less than one hundred words.*

As I said in Section 6.5, mathematicians have therefore adopted a set of rules (typically the Zermelo–Fraenkel axioms together with rules about what constitutes a mathematical statement) governing how we may construct sets and so restricting how we can define mathematical objects. Thus, for example, given sets A and B, the Zermelo–Fraenkel axioms permit us to consider the

---

[4] A set is called countable if it is enumerable and uncountable if not.

product set $A \times B$ consisting of all ordered pairs (a, b) with $a \in A$, $b \in B$ and the *power set* $\mathcal{P}(A)$ consisting of all subsets of A.

When we constructed the strictly positive rationals from the natural numbers, we looked at the set of ordered pairs $(n, n') \in \mathbb{N}^+ \times \mathbb{N}^+$. We then looked at equivalence classes $[(n, n')]$. These are members[5] of $\mathcal{P}(\mathbb{N}^+ \times \mathbb{N}^+)$. The Zermelo–Fraenkel axioms form a padded playpen for mathematicians, permitting us to do what we wish to do and preventing us from doing things, like using self-referential definitions, which we have no wish to do. A good discussion of the Zermelo–Fraenkel axioms will be found in [12].

Although, as the previous paragraph indicates, set theory may form part of the background, it is not necessarily required for the day-to-day study of countable objects like the natural numbers. However, whereas every natural number has a name: one, two, ..., seventeen billion, ... the fact that the reals are uncountable means that 'most real numbers have no name'. (Recall from our discussion of codes in Section 5.3 that we can associate every English sentence with a unique natural number and then every named real number like 'two sevenths' or 'the area of a circle radius one' is associated with a natural number and each natural number is associated with at most one such sentence.) Thus a 'typical real number' forms part of an anonymous mass whose purpose is to form members of sets (like the set of all real numbers x with $0 < x < 1$). Under these circumstances, use of the language of sets becomes essential.

8.3 Are the Real Numbers Real?

To the ancient Egyptians and Greeks, the sun was a unique object. To us, the sun is merely one star among many. As I have remarked several times before, it is much easier to study an object if it is one among many. The real number system seems to play such a unique role in our understanding of the world around us that it is difficult to find anything to compare it with.

However, I think that a comparison with the status of Euclidean geometry may be useful. Until the beginning of the nineteenth century, few people doubted the truth of Euclidean geometry (that is to say, the geometry described in Euclid's *Elements*). Indeed, few people even asked themselves what it meant to say that Euclidean geometry was 'true'. Those philosophers, like Kant, who

---

[5] The alert reader may observe that we have introduced an uncountable set $\mathcal{P}(\mathbb{N}^+ \times \mathbb{N}^+)$. We could have confined ourselves to countable objects here and in the construction of the rationals $\mathbb{Q}$, but this would involve more work for no particular gain.

considered the matter suggested either that space was inherently Euclidean or, more interestingly, that the only way the human mind could understand space was through Euclidean geometry.

Bolyai and Lobachevsky showed that it was possible to develop other axioms to produce different geometries (in particular, ones in which 'the angles of a triangle do not add up to 180 degrees'). Riemann (following his teacher Gauss) approached the matter from a different direction.

If we do geometry on Earth's surface, then lines are replaced by 'great circles' and, although on a sufficiently small scale 'the angles of a triangle formed by great circles add up to 180 degrees to any desired degree of accuracy'[6], this is not true on a large scale, and navigation requires 'spherical geometry' rather than 'flat geometry'. We say that the rules of Euclidean geometry apply locally but that the sphere is 'curved'.

In the picture just given, the spherical surface is pictured as lying in three-dimensional Euclidean space so the 'true geometry is Euclidean', but, when we restrict ourselves to the 'curved surface', it is more convenient to use a non-Euclidean geometry.

Riemann threw away the comfort blanket of a Euclidean space containing our surface and simply considered the surface. Thus, for Riemann, a surface (or *manifold*) was simply something whose local structure was Euclidean, but whose global structure could be very different. Of course, there is no need to restrict ourselves to two-dimensional surfaces. A manifold could be an object that looked locally like *n*-dimensional Euclidean space. Riemann and his successors (including Christoffel, Ricci-Curbastro and Levi-Civita) developed the mathematical methods required by this new conception.

In his theory of special relativity, Einstein replaced the picture of a universe in which the laws of nature conspired (by aether drift or contracting rulers) to make it impossible to measure the differing speeds of light in systems moving with different velocities with respect to each other, by one simple law of nature stating that the velocity of light is constant in all such systems.

A couple of years later, Einstein had what he called the 'happiest thought of [his] life'. In classical physics, the effects of gravitation and acceleration are indistinguishable. Instead of trying to distinguish the indistinguishable, why not treat them as identical? Such a new theory required new mathematics. There is a pleasant story, told by Einstein himself, that he explained his problem to his mathematician friend Grossmann 'and asked him please to go to the library and see if there was an appropriate geometry to handle such questions.

---

[6] The higher the required accuracy, the smaller we must take our triangles.

The next day Grossmann returned ... and said that there was such a geometry, Riemannian geometry'.[7] When, after several years of hard thinking, Einstein completed his general theory of relativity, his new universe was not a surface embedded in some Euclidean space, but a universe described by Riemannian geometry.

Compared with most parts of modern physics, special relativity is mathematically quite simple and experimentally quite easy to test. General relativity is not only mathematically complicated, but also hard to test, since the effects it predicts are small and subtle.

When the first navigational satellite was to be launched, some engineers involved in the project assumed that they would only need to take into account special relativity, while others worried about general relativity. To be on the safe side, a committee was set up to consider the matter. When the committee reported that a general relativistic correction would be required (since, for example, general relativity demands that clocks run faster in lower gravity) the engineers installed a 'general relativity switch' which would be operated if this turned out to be the case. Observations after the launch showed that the committee (and so general relativity) had made the right prediction and the switch was turned on. Without correction, the system would have given positions out by 10 kilometres within a day.[8]

Is it possible that, just as Euclidean geometry turned out not to be the only way of viewing the world and, indeed, not the best way of viewing the world, so the real numbers may not be as fundamental as we think them to be? Since the time of Galileo, physics has viewed the world through 'Stevinian' eyes. The particles of Newtonian mechanics have position and momentum and, by taking sufficient care, we can persuade nature to reveal a further decimal place in their values.

However, quantum mechanics tells us that we cannot know both position and momentum to arbitrary accuracy. If we continue with the Stevinian view that the pair (position, momentum) exists, although we cannot measure it to arbitrary accuracy, are we not repeating the same mistake as those who held that we can only understand nature through Euclidean eyes?

I am not suggesting that modifying our notion of real number is the way to resolve the problems of modern physics. However, it would not surprise me if, in due course, resolving the problems of modern physics led to a change in our notion of number.

---

[7] The story is taken from [26], where it is placed in its proper context.
[8] This account is taken from [1].

And new philosophy calls all in doubt;
The element of fire is quite put out;
The sun is lost, and th' earth, and no man's wit
Can well direct him where to look for it.
And freely men confess that this world's spent,
When in the planets, and the firmament
They seek so many new; they see that this
Is crumbled out again to his atomies.
'Tis all in pieces, all coherence gone,
All just supply, and all relation.

John Donne, An Anatomy of the World

9

The Complex Numbers

9.1 Constructing the Complex Numbers

It is easy to postulate new numbers to 'solve' previously insoluble problems. It is, as we have seen, much harder (and usually impossible) to introduce them in a manner which is consistent with other uses of numbers. And, even when it is possible to introduce them in a consistent manner, it is often not very useful. (See, for example, Exercise 7.2.10.) It is easy to introduce i as a root of $x^2 = -1$, but it is by no means clear why we should do so.

Cardano, who was the first person to explicitly consider the square roots of negative numbers, ended his discussion with the words: 'This subtlety results from arithmetic of which this final point is, as I have said, as subtle as it is useless.' That i might be interesting, at least for those studying mathematical puzzles, is indicated by the formula, familiar to all my readers,

$$\frac{-b \perp \sqrt{b^2 - 4ac}}{2a},$$

which, if we permit the introduction of i, solves all quadratics

$$at^2 + bt + c = 0$$

with $a \neq 0$.

More particularly, if we allow ourselves to consider i as a number, we can factorise a quadratic into linear factors

$$at^2 + bt + c = a(t - \alpha)(t - \beta).$$

Over time, it became clear that a similar thing occurs for cubics and quartics *without the need to introduce any further fictitious numbers*, and there was a strong suspicion that the same might be true for all polynomials.

187

Later, adventurous mathematicians like Leibniz and Euler tried to introduce the standard recipe 'treat fictitious numbers in the same way as ordinary numbers', which had worked so well in algebra, into the new calculus. Some of the formulae which emerged were striking, but others were nonsensical.

Example 9.1.1 *If the reader has done advanced school calculus, she may wish to consider which parts of the argument that follows are definition, which parts are provable, which parts seem plausible and which parts are bluff. Otherwise she can skip the whole thing.*

By using Taylor's theorem, we have

$$e^{ix} \underset{?}{=} \sum_{n=0}^{\infty} \frac{(ix)^n}{n!}$$

$$\underset{?}{=} \sum_{r=0}^{\infty} \frac{(-1)^r x^{2r}}{(2r)!} + i \sum_{r=0}^{\infty} \frac{(-1)^r x^{2r+1}}{(2r+1)!}$$

$$\underset{?}{=} \cos x + i \sin x.$$

In particular, setting $x = \pi$, we have

$$e^{i\pi} \underset{?}{=} -1,$$

so, taking logarithms,

$$\log(-1) \underset{?}{=} i\pi$$

and so

$$0 \underset{?}{=} \log 1 \underset{?}{=} \log\left((-1)^2\right) \underset{?}{=} 2\log(-1) \underset{?}{=} 2\pi i.$$

If complex numbers were to be used in these new ways, they would need to be better understood.

In view of the work we had to do in earlier constructions and the use of words like 'complex' and 'imaginary' in describing the new numbers, the construction of the complex numbers $\mathbb{C}$ from $\mathbb{R}$ turns out to be surprisingly easy.

Definition 9.1.2 *We write $\mathbb{C}$ for the collection of ordered pairs*

$$\mathbf{a} = (a_1, a_2)$$

of real numbers a_1, a_2. We take

$$(a_1, a_2) + (b_1, b_2) = (a_1 + a_2, b_1 + b_2),$$
$$(a_1, a_2) \times (b_1, b_2) = \left((a_1 \times a_2) - (b_1 \times b_2), (a_1 \times b_2) + (a_2 \times b_1)\right).$$

Theorem 9.1.3 *With the definitions just given,* $(\mathbb{C}, +, \times)$ *forms a field with unit* $\mathbf{1} = (1, 0)$ *and zero* $\mathbf{0} = (0, 0)$. *More specifically, whenever* $\mathbf{a}$, $\mathbf{b}$, $\mathbf{c} \in \mathbb{C}$, *we have*

(i) $\mathbf{a} + \mathbf{b} = \mathbf{b} + \mathbf{a}$. *(Commutative law of addition)*

(ii) $\mathbf{a} + (\mathbf{b} + \mathbf{c}) = (\mathbf{a} + \mathbf{b}) + \mathbf{c}$. *(Associative law of addition)*

(iii) $\mathbf{0} + \mathbf{a} = \mathbf{a}$. *(Additive zero)*

(iv) *If we write* $-\mathbf{a} = (-a_1, -a_2)$, *then* $\mathbf{a} + (-\mathbf{a}) = \mathbf{0}$. *(Additive inverse)*

(v) $\mathbf{a} \times \mathbf{b} = \mathbf{b} \times \mathbf{a}$. *(Commutative law of multiplication)*

(vi) $\mathbf{a} \times (\mathbf{b} \times \mathbf{c}) = (\mathbf{a} \times \mathbf{b}) \times \mathbf{c}$. *(Associative law of multiplication)*

(vii) *We have* $\mathbf{1} \times \mathbf{a} = \mathbf{a}$. *(Multiplicative unit)*

(viii) *If* $\mathbf{a} \neq \mathbf{0}$, *then writing*

$$\mathbf{a}^{-1} = \left(\frac{a_1}{a_1^2 + a_2^2}, \frac{-a_2}{a_1^2 + a_2^2} \right)$$

we have $\mathbf{a} \times \mathbf{a}^{-1} = \mathbf{1}$. *(Multiplicative inverse)*

(ix) $\mathbf{a} \times (\mathbf{b} + \mathbf{c}) = (\mathbf{a} \times \mathbf{b}) + (\mathbf{a} \times \mathbf{c})$. *(Distributive law)*

We note that $\mathbf{1} \neq \mathbf{0}$.

Proof. This is all routine verification. We check statements (vi), (viii) and (ix), leaving the rest to the reader.

(vi) Making free use of the laws governing computation in $\mathbb{R}$, we have

$$\mathbf{a} \times (\mathbf{b} \times \mathbf{c}) = (a_1, a_2) \times \big((b_1 \times c_1) - (b_2 \times c_2), (b_1 \times c_2) + (b_2 \times c_1) \big)$$

$$= \Big(\big(a_1 \times ((b_1 \times c_1) - (b_2 \times c_2)) \big) - \big(a_2 \times ((b_1 \times c_2) + (b_2 \times c_1)) \big),$$

$$\big(a_1 \times ((b_1 \times c_2) + (b_2 \times c_1)) \big) + \big(a_2 \times ((b_1 \times c_1) - (b_2 \times c_2)) \big) \Big)$$

$$= \Big(\big(((a_1 \times b_1) - (a_2 \times b_2)) \times c_1 \big) - \big(((a_2 \times b_1) + (a_1 \times b_2)) \times c_2 \big),$$

$$\big(((a_1 \times b_1) - (a_2 \times b_2)) \times c_2 \big) + \big(((a_1 \times b_2) + (a_2 \times b_1)) \times c_1 \big) \Big)$$

$$= (\mathbf{a} \times \mathbf{b}) \times \mathbf{c}.$$

(viii) We have

$$\mathbf{a} \times \mathbf{a}^{-1} = (a_1, a_2) \times \left(\frac{a_1}{a_1^2 + a_2^2}, \frac{-a_2}{a_1^2 + a_2^2} \right)$$

$$= \left(\frac{(a_1 \times a_1) - (a_2 \times -a_2)}{a_1^2 + a_2^2}, \frac{(a_1 \times -a_2) + (a_2 \times a_1)}{a_1^2 + a_2^2} \right)$$

$$= \left(\frac{a_1^2 + a_2^2}{a_1^2 + a_2^2}, \frac{(a_1 \times a_2) - (a_1 \times a_2)}{a_1^2 + a_2^2} \right)$$

$$= (1, 0) = \mathbf{1}.$$

(xi) We have

$$\mathbf{a} \times (\mathbf{b} + \mathbf{c}) = (a_1, a_2) \times (b_1 + c_1, b_2 + c_2)$$
$$= \left((a_1 \times (b_1 + c_1)) - (a_2 \times (b_2 + c_2)), (a_1 \times (b_2 + c_2)) \right.$$
$$\left. + (a_2 \times (b_1 + c_1)) \right)$$
$$= \left(((a_1 \times b_1) - (a_2 \times b_2)) + ((a_1 \times c_1) - (a_2 \times c_2)), \right.$$
$$\left. ((a_1 \times b_2) + (a_2 \times b_1)) + ((a_1 \times c_2) + (a_2 \times c_1)) \right)$$
$$= \left((a_1 \times b_1) - (a_2 \times b_2), (a_1 \times b_2) + (a_2 \times b_1) \right)$$
$$+ \left((a_1 \times c_1) - (a_2 \times c_2), (a_1 \times c_2) + (a_2 \times c_1) \right)$$
$$= (\mathbf{a} \times \mathbf{b}) + (\mathbf{a} \times \mathbf{c}). \qquad \blacksquare$$

Exercise 9.1.4 *Prove as many of the remaining parts of Theorem 9.1.3 as you wish in as much detail as you feel appropriate.*

Check that, writing $i = (0, 1)$, we have $i \times i = -\mathbf{1}$.

Exercise 9.1.5 *Consider the function $f : \mathbb{R} \to \mathbb{C}$ given by $f(x) = (x, 0)$. Show that f is an injection which preserves addition and multiplication. [We develop this idea further in Exercise 9.3.4.]*

The reader will observe that our construction is direct and does not use equivalence classes.[1] From now on, we use the standard notations such as $\mathbf{a} = a_1 + ia_2$, $\mathbf{0} = 0$, $\mathbf{1} = 1$, $z \times w = zw$ and use the various algebraic laws for a field freely. We also make use of Exercise 9.1.5 to consider $\mathbb{R}$ as a subset of $\mathbb{C}$. If $z = x + iy$ with $x, y \in \mathbb{R}$, we refer to x as the *real part* of z and y as the *imaginary part* of z. We write $\mathfrak{R}z = x$ and $\mathfrak{I}z = y$ so that $z = \mathfrak{R}z + i\mathfrak{I}z$.

We have obtained a field in which -1 has a square root. If we were only interested in algebra, there would be little more to be said, but the model of $\mathbb{C}$ (due to Argand) is much more interesting than that. From our point of view, what Argand did is to equip a geometrical object (the two-dimensional

---

[1] I give a construction using equivalence classes in Exercise 10.4.20 together with an explanation of why I prefer the one given here.

(x, y) plane familiar from Cartesian geometry) with an algebraic structure.[2] This means that we can use our geometric intuition to guide our algebraic steps.

In fact, as Cauchy discovered, even more is true; we can do analysis with $\mathbb{C}$ in a way which reflects the geometry of the plane. Cauchy's *complex analysis*[3] has been a major subject of study for the last two centuries and finds uses everywhere from number theory through the stability of electronic devices to the central limit theorem of probability.

We will not treat complex analysis in this book, but in the next section will show how the idea of a continuous function can be generalised to this new context. Since we have laid so much emphasis on *ordered fields* and *bounded increasing sequences*, it is by no means clear how we shall deal with $\mathbb{C}$, which *cannot be ordered.*

Exercise 9.1.6 *Suppose that* $\mathbb{F}$ *is an ordered field. Show that the equation* $x^2 = -1$ *has no solution.*
[See Exercise 3.2.16 if you need a hint.]

9.2 Analysis for $\mathbb{C}$

The key to the extension of calculus to more general objects than $\mathbb{R}$ is the notion of distance. We write X^2 for the set of ordered pairs (a, b) with a, $b \in X$.

Definition 9.2.1 *Let* X *be a non-empty set. We call a function* $d : X^2 \to \mathbb{R}$ *a metric (or distance function) if the following conditions hold for every* $a, b, c \in X$.

(i) $d(a, b) \geq 0$. *(The distance between two points* a *and* b *is always positive.)*

(ii) $d(a, b) = 0$ *if and only if* $a = b$. *(The distance between a point and itself is zero and, if two points are at zero distance, they are identical.)*

(iii) $d(a, b) = d(b, a)$. *(The distance from* a *to* b *is the same as the distance from* b *to* a.)

(iv) *(Triangle inequality)* $d(a, b) + d(b, c) \geq d(a, c)$. *(The distance from* a *to* c *is no greater than the distance from* a *to* c *via* b.)

---

[2] From the point of view of his contemporaries, Argand gave an algebraic object a geometric interpretation. Argand published his idea as an anonymous booklet and the first paper to discuss it ends with an appeal to the author of this booklet to make himself known.

[3] Another subject with lots of results named after Cauchy.

Our distance will be derived from the *modulus function*

$$|x + iy| = \sqrt{x^2 + y^2}$$

for x, $y \in \mathbb{R}$. (Note that the existence and uniqueness of the square root is guaranteed by Lemma 7.4.12.)

Lemma 9.2.2

(i) If z, $w \in \mathbb{C}$, then $|zw| = |z||w|$.
(ii) If we write $d(z, w) = |z - w|$, then d is a metric on $\mathbb{C}$.

Lemma 9.2.2 can be easily verified by direct calculation. I give a slightly more subtle proof in the next exercise, which uses the notion of the *complex conjugate*.

Definition 9.2.3 *We work in $\mathbb{C}$. If x and y are real, we write*

$$(x + iy)^* = x - iy.$$

We say that z^* is the complex conjugate *of z.*

Exercise 9.2.4 *We take z, $w \in \mathbb{C}$.*

(i) Show that $(z^*)^* = z$, $|z^*| = |z|$, $(z + w)^* = z^* + w^*$ and $(zw)^* = z^*w^*$.
(ii) Show that $zz^* = |z|^2$.
(iii) By considering $(zw)(zw)^*$, show that $|zw| = |z||w|$.
(iv) Show that, if x, $y \in \mathbb{R}$, then $(x+iy)+(x+iy)^* = 2x$. Deduce that $z + z^*$ is real and $z + z^* \leq 2|z|$.
(v) By considering $(z + w)(z + w)^*$, show that $|z + w| \leq |z| + |w|$.
(vi) Show that $d(z, w) = |z - w|$ defines a metric on $\mathbb{C}$.

[*We remark that (i) shows that the function $f : \mathbb{C} \to \mathbb{C}$ defined by $f(z) = z^*$ is a field isomorphism (see Definition 5.1.5) of $\mathbb{C}$ with itself. The existence of this isomorphism reflects the fact that i and $-i$ have equal status as square roots of -1.*]

We make the following easy, but necessary, remark.

Exercise 9.2.5

(i) We have already defined the modulus $|x|_{\mathbb{R}}$ for x real (see Exercise 7.2.4). We have now defined the modulus $|z|_{\mathbb{C}}$ for z complex. Show that

$$|x + i0|_{\mathbb{C}} = |x|_{\mathbb{R}}$$

for all x real. Thus the modulus function is backwards compatible and we
can write $|\ |$ *rather than* $|\ |_\mathbb{R}$ *or* $|\ |_\mathbb{C}$.
(ii) If x and y are real, show that

$$|x|, \ |y| \le |x + iy| \le |x| + |y|.$$

We define the limit of a sequence in $\mathbb{C}$ in the same way that we did for
ordered fields.

Definition 9.2.6 *We work in* $\mathbb{C}$. *We say that* $z_n \to z$ *(or, in words,* z_n *tends to a*
limit z) if, given $\epsilon > 0$, *we can find an N such that* $|z - z_n| < \epsilon$ *for all* $n \ge N$.

Notice, however, that we have a vivid geometric picture of points z_n getting closer to a point z in the complex plane. The proofs of the elementary
properties of limits are the same as we gave in Lemma 7.3.3 and are left as an
exercise.

Exercise 9.2.7 *We use the notations and assumptions of Definition 9.2.6.*
Prove the following results.

 (i) If z_n *tends to a limit, then that limit is unique.*
 (ii) If $z_n \to z$ *and* $w_n \to w$, *then* $z_n + w_n \to z + w$ *as* $n \to \infty$.
 (iii) If $z_n \to z$ *and* $w_n \to w$, *then* $z_n w_n \to zw$ *as* $n \to \infty$.
 (iv) If $z_n = z$ *for all n, then* $z_n \to z$ *as* $n \to \infty$.

Exercise 9.2.8 *Prove the following results.*

 (i) $\big||a| - |b|\big| \le |a - b|$ *for a, b* $\in \mathbb{C}$.
 (ii) If $z_n \to z$ *in* $\mathbb{C}$ *as* $n \to \infty$, *then* $|z_n| \to |z|$ *in* $\mathbb{R}$.
 (iii) If $z_n \to z$ *in* $\mathbb{C}$ *as* $n \to \infty$ *and* $|z_n| \le A$ *for all n, then* $|z| \le A$.

Exercise 9.2.9 *Suppose that* $z_n = x_n + iy_n$, $z = x + iy$ *with* x_n, y_n, x, y *real.*
Show that $z_n \to z$ *in* $\mathbb{C}$ *if and only if* $x_n \to x$ *and* $y_n \to y$ *in* $\mathbb{R}$ *as* $n \to \infty$.
Show also that, if $z_n \to z$, *then* $z_n^* \to z^*$ *as* $n \to \infty$.

In Section 7.6 we discussed various equivalent forms of the fundamental
axiom. The fundamental axiom itself and the supremum principle make direct
use of the fact that $\mathbb{R}$ is ordered, so we cannot expect them to extend to $\mathbb{C}$.
However, the Bolzano–Weierstrass property and the notion of completeness
can be extended.

We start with the Bolzano–Weierstrass property.[4]

---

[4] In more advanced work the Bolzano–Weierstrass property is called *sequential compactness*.

Theorem 9.2.10 **(Bolzano–Weierstrass for** $\mathbb{C}$**)** *Let R be a fixed strictly positive real number. If $z_n \in \mathbb{C}$ and $|z_n| \le R$ for each n, then we can find $1 \le j(1) < j(2) < j(3) < \cdots$ and $z \in \mathbb{C}$ such that $z_{j(n)} \to z$ as $n \to \infty$.*

Proof. Write $z_n = x_n + iy_n$ with x_n and y_n real. We have $|x_n| \le |z_n| \le R$ so, by the Bolzano–Weierstrass property of $\mathbb{R}$, we can find $x \in \mathbb{R}$ and integers $1 \le k(1) < k(2) < \cdots$ such that $x_{k(n)} \to x$ in $\mathbb{R}$. We now have $|y_{k(n)}| \le |z_{k(n)}| \le R$ so, again, by the Bolzano–Weierstrass property of $\mathbb{R}$, we can find $y \in \mathbb{R}$ and integers $1 \le l(1) < l(2) < \cdots$ such that $y_{k(l(n)))} \to y$ in $\mathbb{R}$. If we now set $j(n) = k(l(n))$ we see that

$$1 \le j(1) < j(2) < \cdots \text{ and } x_{j(n)} \to x, \ y_{j(n)} \to y \text{ in } \mathbb{R} \text{ as } n \to \infty.$$

Thus, writing $z = x + iy$, we have

$$|z - z_{j(n)}| \le |x - x_{j(n)}| + |y - y_{j(n)}| \to 0 + 0 = 0$$

as $n \to \infty$, so we are done. ∎

We next extend the notion of Cauchy sequence and completeness. As before, the reader should note that, although the words remain the same, the geometric picture becomes richer.

Definition 9.2.11 *We work in $\mathbb{C}$. We say that a sequence (z_n) is a Cauchy sequence if, given $\epsilon > 0$, we can find an N such that $|z_n - z_m| < \epsilon$ for all $n, m \ge N$.*

Exercise 9.2.12 *We work in $\mathbb{C}$. Show that every convergent series is Cauchy.*

Theorem 9.2.13 *Every Cauchy sequence in $\mathbb{C}$ converges.*

I give two alternative proofs as exercises. Both provide useful practice in working with $\mathbb{C}$.

Our first proof (like our proof of Theorem 9.2.10) depends on splitting into real and imaginary parts.

Exercise 9.2.14 *Suppose that we have a Cauchy sequence (z_n) in $\mathbb{C}$. Let us write $z_n = x_n + iy_n$ with $x_n, y_n \in \mathbb{R}$.*

(i) Show that x_n and y_n form Cauchy sequences in $\mathbb{R}$.

(ii) Deduce that we can find $x, y \in \mathbb{R}$ such that $x_n \to x$ and $y_n \to y$ as $n \to \infty$.

(iii) Set $z = x + iy$ and show that $z_n \to z$ as $n \to \infty$.

Our second proof is considerably longer, but, by avoiding the special technique of splitting into real and imaginary parts, it gives greater promise of generalisation.

Exercise 9.2.15

(i) Reread the paragraph on page 164 beginning 'Suppose that we have a Cauchy sequence (a_n)'.

(ii) If (z_n) is a Cauchy sequence in $\mathbb{C}$, show that there exists an R with $|z_n| \leq R$ for all n.

(iii) Deduce from the Bolzano–Weierstrass theorem for $\mathbb{C}$ that the sequence (z_n) has a convergent subsequence.

(iv) Conclude that the sequence (z_n) converges.

9.3 Continuous Functions from $\mathbb{C}$

We can define continuous functions from the complex plane in much the same way as we defined continuous functions on $\mathbb{R}$.

Definition 9.3.1 Let $\mathbb{F} = \mathbb{R}$ or $\mathbb{F} = \mathbb{C}$ and let $f : \mathbb{C} \to \mathbb{F}$ be a function. We say that f is continuous at $z \in \mathbb{C}$ if, whenever $z_n \to z$, we have $f(z_n) \to f(z)$. We say that f is continuous if it is continuous at every point z of $\mathbb{C}$.

Exercise 9.3.2 Let $\mathbb{F} = \mathbb{R}$ or $\mathbb{F} = \mathbb{C}$. Show that, if $f, g : \mathbb{C} \to \mathbb{F}$ are continuous functions, then the sum function $h = f + g$, defined by $h(z) = f(z) + g(z)$ for all $z \in \mathbb{C}$, and the product function $k = f \times g$, defined by $k(z) = f(z) \times g(z)$ for all $z \in \mathbb{C}$, are continuous.

Exercise 9.3.3 Show that, if $f : \mathbb{C} \to \mathbb{C}$ is continuous, then, writing $g(z) = |f(z)|$ we have $g : \mathbb{C} \to \mathbb{R}$ continuous.

Exercise 9.3.4 Write down the appropriate definition for a continuous function $h : \mathbb{R} \to \mathbb{C}$.

Consider the function $f : \mathbb{R} \to \mathbb{C}$ given by $f(x) = (x, 0)$ (where we revert to the notation $(x, y) = x + iy$). Show that f is a continuous injection which preserves addition and multiplication.

Show that if $g : \mathbb{R} \to \mathbb{C}$ is continuous function which preserves addition and multiplication and $g(t) \neq 0$ for some $t \in \mathbb{R}$, then $f(x) = g(x)$ for all x.

If the reader reflects, she will see that it is by no means clear what sort of theorem would correspond to the intermediate value theorem when we consider general continuous functions $f : \mathbb{C} \to \mathbb{C}$.

However, we can prove another powerful theorem.

Theorem 9.3.5 *Suppose that* $f : \mathbb{C} \to \mathbb{C}$ *is continuous and* $R > 0$. *Then we can find* w, $w' \in \mathbb{C}$ *such that* $|w|$, $|w'| \leq R$ *and*

$$|f(w')| \leq |f(z)| \leq |f(w)|$$

for all z *with* $|z| \leq R$.

The reader who doubts the usefulness of this should have her doubts removed in Section 10.2. The reader who thinks the theorem is obvious should look at Exercise 9.3.9.

We shall prove a very slightly stronger theorem.

Theorem 9.3.6 *Suppose that* $g : \mathbb{C} \to \mathbb{R}$ *is continuous and* $R > 0$. *Then we can find* w, $w' \in \mathbb{C}$ *such that* $|w|$, $|w'| \leq R$ *and*

$$g(w') \leq g(z) \leq g(w)$$

for all z *with* $|z| \leq R$.

Proof of Theorem 9.3.5 from Theorem 9.3.6. Write $g(z) = |f(z)|$ and observe that $g : \mathbb{C} \to \mathbb{R}$ is continuous.   ∎

Our proof of Theorem 9.3.6 passes through a preliminary result.

Theorem 9.3.7 *Suppose that* $g : \mathbb{C} \to \mathbb{R}$ *is a continuous function. Then, if* $R > 0$, *we can find a* K *and* K' *such that*

$$K' \leq g(z) \leq K$$

for all z *with* $|z| \leq R$.

Proof. We prove the existence of K. The existence of K' follows by a similar argument or by considering $-g$.

Suppose, if possible, that $g : \mathbb{C} \to \mathbb{R}$ is a continuous function and $R > 0$, but we cannot find a K such that $g(z) \leq K$ for all z with $|z| \leq R$.

Set $w_1 = 0$ and construct a sequence w_n as follows. Given w_n our hypothesis tells us that it is not true that $g(z) \leq g(w_n) + 1$ for all z with $|z| \leq R$. Thus we may choose w_{n+1} with $|w_{n+1}| \leq R$, but $g(w_{n+1}) \geq g(w_n) + 1$.

By the Bolzano–Weierstrass theorem for $\mathbb{C}$ (Theorem 9.2.10), we can find a $w \in \mathbb{C}$ and a sequence $n(1) < n(2) < \cdots$ such that $w_{n(j)} \to w$ as $j \to \infty$. By the continuity of g, this implies that $g(w_{n(j)}) \to g(w)$ so

$$|g(w_{n(j)}) - g(w_{n(j+1)})| \leq |g(w_{n(j)}) - g(w)| + |g(w_{n(j+1)}) - g(w)| \to 0,$$

and so $g(w_{j(n)}) \to g(w)$ as $j \to \infty$. By construction, $g(w_{n(j)}) + 1 \leq g(w_{n(j+1)})$, so, taking limits, $g(w) + 1 \leq g(w)$ and we have a contradiction. ∎

Proof of Theorem 9.3.6. We prove the existence of w. The proof of the existence of w' follows by a similar argument or by considering $-g$.

By Theorem 9.3.7, the set E of $g(z)$ with $|z| \leq R$ forms a non-empty subset of $\mathbb{R}$ which is bounded above. By the supremum property of $\mathbb{R}$, it follows that E has a supremum, that is to say, there exists an $a \in \mathbb{R}$ with the following properties.

(1) $a \geq g(z)$ for all z with $|z| \leq R$.

(2) If $b \geq g(z)$ for all z with $|z| \leq R$, then $b \geq a$.

By condition (2), there exist z_n with $|z_n| \leq R$ such that $g(z_n) \geq a - 1/n$. Using (1), we see that $a - 1/n \leq g(z_n) \leq a$ and so, by the axiom of Archimedes, $g(z_n) \to a$. By the Bolzano–Weierstrass theorem for $\mathbb{C}$, we can find a $w \in \mathbb{C}$ and a sequence $n(1) < n(2) < \cdots$ such that $z_{n(j)} \to w$ as $j \to \infty$. By the continuity of g, we have $g(z_{n(j)}) \to g(w)$. By the uniqueness of limits $g(w) = g(a)$. By (1), $g(a) \geq g(z)$ for all z with $|z| \leq R$.

Finally we note that $|z_{n(j)}| \leq R$ and $|z_{n(j)}| \to a$ so $|a| \leq R$ and we are done ∎

In Appendix B we shall need a minor variation on Theorem 9.3.7.

Exercise 9.3.8 *Suppose that $g : \mathbb{C} \to \mathbb{R}$ is continuous and $R > 0$. Show that there exist w, $w' \in \mathbb{C}$ such that $|w|$, $|w'| = R$ and*

$$g(w') \leq g(z) \leq g(w)$$

for all z with $|z| = R$.

Exercise 9.3.9 *Consider* $\mathbb{A}$ *consisting of those $z \in \mathbb{C}$ of the form $z = a + bi$ with a, $b \in \mathbb{Q}$.*

(i) *Show that, if z_1, $z_2 \in \mathbb{A}$, then $z_1 + z_2$, $z_1 \times z_2$, $-z_1$, $z_1^* \in \mathbb{A}$. Show that, if $z_1 \in \mathbb{A}$ and $z_1 \neq 0$, then $z_1^{-1} \in \mathbb{A}$. Conclude that $\mathbb{A}$ with the operations inherited from $\mathbb{C}$ is a field.*

(iii) *Show that $2 \in \mathbb{A}$, but there is no $w \in \mathbb{A}$ with $w^2 - 2 = 0$.*

(vi) *Show that $2 \in \mathbb{A}$, but there is no $w \in \mathbb{A}$ with $|w^2 - 2| \leq |z^2 - 2|$ for all $z \in \mathbb{A}$.*

10

A Plethora of Polynomials

10.1 Preliminaries

We have completed our construction of number systems starting from the Peano axioms and ending with the complex numbers.

However, we extended the rationals to the reals not for the sake of the reals themselves, but for the sake of studying functions $f : \mathbb{R} \to \mathbb{R}$. The standard first course in analysis thus consists of the major theorems (the intermediate value theorem, the fact that a continuous function on a closed bounded interval attains its bounds[1] and the mean value theorem), followed by methods such as power series and differential equations for constructing a wide range of interesting functions.

The only general class of functions we have immediately to hand are the polynomials and we shall use them in this penultimate chapter to illustrate some points raised earlier.

In this first section we establish some properties of polynomials. Since most of these are already familiar to the reader and since we wish to use, rather than discuss, these properties, the reader should feel free to skim or, indeed, skip these preliminaries.

If the reader reflects, she will see that the polynomials are, in effect, defined inductively.

Definition 10.1.1 *We work in a field $\mathbb{F}$. We say that $P : \mathbb{F} \to \mathbb{F}$ is a polynomial of degree at most 0 if there is an $a \in \mathbb{F}$ such that $P(u) = a$ for all u. If $a = 0$ we say that P is the zero polynomial. If $a \neq 0$, we say that P has degree 0 and leading coefficient a. If $n \geq 1$, we say that $P : \mathbb{F} \to \mathbb{F}$ is a polynomial of degree at most n if $P(u) = au^n + Q(u)$ for all $u \in \mathbb{F}$ where $a \in \mathbb{F}$ and Q is a polynomial of degree at most $n - 1$. If $a \neq 0$ in the previous sentence, we say*

---

[1] A variation on the theme of Theorem 9.3.7.

*that P is a polynomial of degree n with leading coefficient a. We say that the
zero polynomial has leading coefficient 0.*

(There is no particular advantage in defining the degree of the zero polynomial. If we talk about a polynomial of degree n we shall always take
$n \geq 0$.)

Although our definition and many of the results in this section work for a
general field $\mathbb{F}$, the reader is advised to stick to $\mathbb{F} = \mathbb{R}$, $\mathbb{Q}$ or $\mathbb{C}$, particularly in
view of the next exercise.

Exercise 10.1.2 *We work in $\mathbb{Z}_2$. Check that if $P(u) = u^2 + u$, then $P(u) = 0$
for all $u \in \mathbb{Z}_2$.*

*Show that if $Q(u) = u^2$ and $R(u) = u$, then $Q(u) = R(u)$ for all $u \in \mathbb{Z}_2$.
Thus two polynomials Q and R of different degree can take the same values.
[In Exercise 10.1.10 we show that this phenomenon does not occur for $\mathbb{R}$, $\mathbb{Q}$
or $\mathbb{C}$. More precisely, we show that, for these fields, a polynomial of degree at
least zero cannot vanish everywhere. The successful study of polynomials over
fields like $\mathbb{Z}_2$ requires algebraists to consider polynomials as 'defined by their
coefficients rather than their values', but that is another story.]*

We can establish properties of polynomials by repeated (and rather routine)
use of induction.

Exercise 10.1.3 *We use the notation of Definition 10.1.1. Prove whichever of
the following results you wish.*

(i) *A polynomial of degree at most n is either the zero polynomial or a
polynomial of degree r for some r with $0 \leq r \leq n$.*

(ii) *Suppose that P is a polynomial of degree n with leading coefficient a and
R is a polynomial of degree m with leading coefficient b. If $n > m \geq 0$,
then $P + R$ is a polynomial of degree n with leading coefficient a. If
$n = m \geq 0$ and $a + b \neq 0$, then $P + R$ is a polynomial of degree n with
leading coefficient $a + b$. If $n = m \geq 1$ and $a + b = 0$, then $P + R$ is a
polynomial of degree at most $n - 1$.*

(iii) *If $c \in \mathbb{F}$, $c \neq 0$ and P is a polynomial of degree n with leading coefficient
a, then the function $R = cP$ (defined by $R(u) = cP(u)$) is a polynomial
of degree n with leading coefficient ca.*

(iv) *If P is a polynomial of degree n with leading coefficient a, then the function R defined by $R(u) = uP(u)$ is a polynomial of degree $n + 1$ with
leading coefficient a.*

(v) *If $b \in \mathbb{F}$, then the formula $P_n(u) = (u - b)^n$ defines a polynomial of degree n.*

(vi) *If $b \in \mathbb{F}$ and P is a polynomial of degree n, then the function R defined by $R(u) = P(u - b)$ is a polynomial of degree n.*

(vii) *If P is a polynomial of degree n with leading coefficient a and Q is a polynomial of degree m with leading coefficient b, then $P \times Q$, defined by the formula $(P \times Q)(u) = P(u)Q(u)$, is a polynomial of degree $n + m$ with leading coefficient ab.*

We now introduce long division. (Compare Lemma 4.3.4.)

Lemma 10.1.4 *If P is a polynomial of degree n with $n \geq 1$ and $b \in \mathbb{F}$, then we can find an $r \in \mathbb{F}$ and a polynomial Q of degree $n - 1$ such that*

$$P(u) = (u - b)Q(u) + r$$

for all $u \in \mathbb{F}$.

Proof. We use induction. If $n = 1$, then, by definition,

$$P(u) = au + d$$

for some $a, d \in \mathbb{F}$ with $a \neq 0$. Thus

$$P(u) = (u - b)a + r$$

with $r = ab + d$.

Suppose now that the result is true for all $n \leq m$ and P is a polynomial of degree $m + 1$. We then have

$$P(u) = au^{m+1} + R(u)$$

for some $a \neq 0$ and some polynomial R of degree at most m. Thus

$$P(u) = a(u - b)u^m + R_1(u),$$

where $R_1(u) = abu^m + R(u)$ and so R_1 is a polynomial of degree at most m. If R_1 is a constant function, we are done. Otherwise, by the inductive hypothesis, we can find a polynomial Q_1 of degree at most $m - 1$ and an $r \in \mathbb{F}$ such that

$$R_1(u) = (u - b)Q_1(u) + r.$$

Thus

$$P(u) = a(u - b)u^m + (u - b)Q_1(u) + r = (u - b)(au^m + Q_1(u)) + r.$$

Since $Q(u) = au^m + Q_1(u)$ defines a polynomial of degree m (by various parts of Exercise 10.1.3), we are done. ∎

A simple remark allows us to improve Lemma 10.1.4.

Theorem 10.1.5 *If P is a polynomial of degree $n \geq 1$ and $b \in \mathbb{F}$, then we can find a polynomial Q of degree $n - 1$ such that*

$$P(u) = (u - b)Q(u) + P(b)$$

for all $u \in \mathbb{F}$.

Proof. If $P(u) = (u - b)Q(u) + r$ for all $u \in \mathbb{F}$, then, setting $u = b$, we obtain $P(b) = r$. ∎

Theorem 10.1.5 has various useful consequences.

Exercise 10.1.6 *Give one line proofs of the following statements about a polynomial P.*

(i) If P has degree $n \geq 1$ and $P(a) = 0$, then

$$P(u) = (u - a)Q(u),$$

where Q is a polynomial of degree $n - 1$.
(ii) If P has degree $n \geq 1$ and $a \in \mathbb{F}$, then

$$P(u) - P(a) = (u - a)Q(u),$$

where Q is a polynomial of degree $n - 1$.

Theorem 10.1.7 *If P is a polynomial of degree at most n and we can find distinct $a_1, a_2, \ldots, a_{n+1} \in \mathbb{F}$ such that $P(a_j) = 0$ for $1 \leq j \leq n + 1$, then $P = 0$. (In other words, if a polynomial of degree at most n vanishes at $n + 1$ points, then P vanishes everywhere. The result is sometimes stated as 'a polynomial of degree n can have at most n roots'.)*

Proof. The result is immediate if $n = 0$. Suppose that it is true for $n = m$ and P is a polynomial of degree at most $m + 1$ such that we can find distinct elements $a_1, a_2, \ldots, a_{m+2} \in \mathbb{F}$ with $P(a_j) = 0$ for $1 \leq j \leq m + 2$. Since $P(a_{m+2}) = 0$, Exercise 10.1.6, tells us that

$$P(u) = (u - a_{m+2})Q(u)$$

for all $u \in \mathbb{F}$ where Q is a polynomial of degree at most m. Now $a_j - a_{m+2} \neq 0$ and

$$0 = (a_j - a_{m+2})Q(a_j)$$

so $Q(a_j) = 0$ for $1 \leq j \leq m+1$. Our inductive hypothesis tells us that $Q = 0$ and so $P = 0$. ∎

Exercise 10.1.8 *If P and Q are polynomials of degree at most n and we can find distinct $a_1, a_2, \ldots, a_{n+1} \in \mathbb{F}$ such that $P(a_j) = Q(a_j)$ for $1 \leq j \leq n + 1$, show that $P(u) = Q(u)$ for all $u \in \mathbb{F}$.*

Exercise 10.1.9 *Suppose that P is a non-zero polynomial of degree at most n such that $P(a) = 0$ for some $a \in \mathbb{F}$. Show that there is an integer m with $n \geq m \geq 1$ and a polynomial Q of degree at most $n - m$ such that $Q(a) \neq 0$ and*

$$P(u) = (u - a)^m Q(u)$$

for all $u \in \mathbb{F}$.

Exercise 10.1.10

(i) *Suppose that $\mathbb{F}$ is infinite. Use Theorem 10.1.7 to show that, if P is a polynomial of degree at least 0, then there exists a $u \in \mathbb{F}$ such that $P(u) \neq 0$.*
(ii) *Explain why your argument for part (i) fails for the example given in Exercise 10.1.2.*

We now specialise to the cases $\mathbb{F} = \mathbb{R}$ or $\mathbb{F} = \mathbb{C}$.

Exercise 10.1.11 *Let $\mathbb{F} = \mathbb{R}$ or $\mathbb{F} = \mathbb{C}$.*

(i) *Use Exercise 7.3.6 or Exercise 9.3.2 and induction to show that the function $f_n : \mathbb{F} \to \mathbb{F}$ given by $f_n(u) = u^n$ is continuous for all $n \geq 1$.*
(ii) *Show that polynomials are continuous.*

We shall need results on the growth of $|P(u)|$ when $|u|$ becomes large.

Lemma 10.1.12 *Let $\mathbb{F} = \mathbb{R}$ or $\mathbb{F} = \mathbb{C}$. If P is a polynomial of degree n (where $n \geq 0$), then we can find real numbers A, B, $R > 0$ (all depending on P) such that*

$$A|u|^n \geq |P(u)| \geq B|u|^n$$

whenever $|u| \geq R$.

Proof. As usual, we prove the result by induction. The result is trivial when $n = 0$, for then $P(u) = a$ for some $a \neq 0$ and all u, so we may take $A = B = |a|$ and $R = 1$.

Now suppose the result is true for all polynomials of degree m or less. If P is a polynomial of degree $m + 1$, we know that

$$P(u) = au^{m+1} + Q(u)$$

where $a \neq 0$ and Q is a polynomial of degree at most m. If $Q = 0$, we may take $R = 1$ and $A = B = |a|$. Otherwise, Q is a polynomial of degree k with $m \geq k \geq 0$. By the inductive hypothesis, we can find A', B', $R' > 0$ (all depending on Q) such that

$$A'|u|^k \geq |Q(u)| \geq B'|u|^k$$

for $|a| \geq R'$. We may certainly assume, in addition, that $R' \geq 1$.

We first observe that, provided $|u| \geq R'$,

$$|P(u)| \leq |a||u|^{m+1} + A'|u|^k \leq |a||u|^{m+1} + A'|u|^{m+1} = (|a| + A')|u|^{m+1}.$$

We also note that, provided $|u| \geq R'$,

$$|P(u)| \geq |a||u|^{m+1} - |Q(u)| \geq |a||u|^{m+1} - A'|u|^k$$
$$= |a||u|^{m+1} - A'|u|^m = |u|^m(|a||u| - A').$$

Thus, if $|u| \geq \max\{R', 2A'/|a|\}$,

$$|P(u)| \geq |u|^m(|a||u|/2) = |a||u|^{m+1}/2.$$

Taking $A = |a| + A'$, $B = |a|/2$ and choosing R so that $R \geq R'$ and $R \geq 2A'/|a|$, we have

$$A|u|^{m+1} \geq |P(u)| \geq B|u|^{m+1}$$

whenever $|u| \geq R$. The required result now follows by induction. ■

Exercise 10.1.13 *Let $\mathbb{F} = \mathbb{R}$. Suppose P is a polynomial of degree n with n odd. Show that we can find a and b such that $P(a) < 0 < P(b)$. Deduce, using the intermediate value theorem, that there exists a c with $P(c) = 0$*

We shall also need the following result, for which the reader may already know a number of proofs but which can also be proved directly by induction.

Exercise 10.1.14 *Let $\mathbb{F} = \mathbb{R}$ or $\mathbb{F} = \mathbb{C}$. For each integer $n \geq 1$, we can find an $A_n > 0$ and a $\delta_n > 0$ such that*

$$|(1 + u)^n - 1 - nu| \leq A_n|u|^2$$

for all $u \in \mathbb{F}$ with $|u| \leq \delta_n$.

10.2 The Fundamental Theorem of Algebra

Much of the work on extending the number system from the strictly positive rationals to the rationals, from the rationals to the reals and from the reals to the complex numbers was inspired by the desire to provide solutions first for linear, then for quadratic and then for more general polynomial equations. The fundamental theorem of algebra shows that this programme was successful.

Theorem 10.2.1 **(Fundamental theorem of algebra)** *If we work over* $\mathbb{C}$, *then, if P is a polynomial of degree n with* $n \geq 1$, *there exists an* $\alpha \in \mathbb{C}$ *with* $P(\alpha) = 0$.

It is, I think, no coincidence that the first mathematicians to produce acceptable proofs of the fundamental theorem[2] were Argand and Gauss, who grasped the *geometric* meaning of $\mathbb{C}$.

It should be noted that the fundamental theorem of algebra is a theorem of *analysis*, since it ultimately depends on the fundamental axiom of analysis for $\mathbb{R}$. [Exercise 9.3.9 (iii) illustrates this.] It should also be noticed that although the fundamental theorem tells us that roots exist it does not give a method for finding them.[3]

There are many proofs of the fundamental theorem of algebra. The one we shall give goes back to Argand's original proof and depends on Theorem 9.3.5 to tell us that a certain function attains a minimum.

By the time students meet the fundamental theorem of algebra in a standard university course, they will know a geometric proof (depending on the notion of angle) for the following very special case of the fundamental theorem.

Lemma 10.2.2 *If* $a \in \mathbb{C}$, $a \neq 0$ *and m is an integer with* $m \geq 1$, *then the equation*

$$az^m = |a|$$

has a solution.

In order to make the book self-contained, I include a proof of Lemma 10.2.2 in Appendix B which does not depend on the idea of angle. However, the reader should feel free to ignore this and should certainly concentrate on understanding the rest of this section before looking at the appendix.

---

[2] The word 'acceptable' complicates matters. We shall return to this point at the end of the section.

[3] The search for such a method is another story, indeed, at least two other stories.

Returning to our main theme, we observe that the fundamental theorem of algebra (Theorem 10.2.1) will follow if we can prove the following two complementary results.

Theorem 10.2.3 *If P is a polynomial of degree n with n $\geq$ 1, there exists an $\alpha \in \mathbb{C}$ with $|P(\alpha)| \leq |P(z)|$ for all $z \in \mathbb{C}$.*

Theorem 10.2.4 (Argand's lemma) *If P is a polynomial of degree n with n $\geq$ 1, and $\alpha \in \mathbb{C}$ has the property that $|P(\alpha)| \leq |P(z)|$ for all $z \in \mathbb{C}$, then $P(\alpha) = 0$.*

Theorem 10.2.3 is a consequence of Theorem 9.3.5.

Proof of Theorem 10.2.3. By Lemma 10.1.12, we can find real numbers $B > 0$ and $R' > 0$ such that

$$|P(z)| \geq B|z|^n$$

whenever $|z| \geq R'$. Now choose $R \geq R'$ such that $BR^n > |P(0)|$. Automatically,

$$|P(z)| \geq B|z|^n \geq BR^n > |P(0)|$$

for all $z \in \mathbb{C}$ with $|z| \geq R$.

By Theorem 9.3.5, we can find $\alpha \in \mathbb{C}$ such that $|\alpha| \leq R$ and

$$|P(z)| \geq |P(\alpha)|$$

for all $z \in \mathbb{C}$ with $|z| \leq R$. Automatically, $|P(0)| \geq |P(\alpha)|$, and the result of the previous paragraph tells us that $|P(z)| > |P(0)|$ for all $z \in \mathbb{C}$ with $|z| \geq R$. It follows that

$$|P(z)| \geq |P(\alpha)|$$

for all z and we are done. ∎

In order to prove Argand's lemma we make a series of simplifications.

Lemma 10.2.5 *The following statements are equivalent.*

(i) If P is a polynomial of degree n with n $\geq$ 1 and $\alpha \in \mathbb{C}$ has the property that $|P(\alpha)| \leq |P(z)|$ for all $z \in \mathbb{C}$, then $P(\alpha) = 0$.

(ii) If P is a polynomial of degree n with n $\geq$ 1 which has the property that $|P(0)| \leq |P(z)|$ for all $z \in \mathbb{C}$, then $P(0) = 0$.

(iii) If P is a polynomial of degree n with n $\geq$ 1 such that $P(0)$ is real, $P(0) \geq 0$ and $|P(0)| \leq |P(z)|$ for all $z \in \mathbb{C}$, then $P(0) = 0$.

(iv) *Suppose that P is a polynomial of the form $P(z) = a + z^m Q(z)$ with a real, $a \geq 0$, m an integer, $m \geq 1$ and Q a polynomial with $Q(0) \neq 0$. If $|P(0)| \leq |P(z)|$ for all $z \in \mathbb{C}$, then $P(0) = 0$.*

(v) *Suppose that P is a polynomial of the form $P(z) = a + z^m Q(z)$ with a real, $a \geq 0$, m an integer, $m \geq 1$ and Q a polynomial with $Q(0)$ real and $Q(0) < 0$. If $|P(0)| \leq |P(z)|$ for all $z \in \mathbb{C}$, then $P(0) = 0$.*

(vi) *Suppose that P is a polynomial of the form $P(z) = a - bz^m + z^{m+1} V(z)$ with a, b real, $a \geq 0$, $b > 0$, m an integer, $m \geq 1$ and V a polynomial. If $|P(0)| \leq |P(z)|$ for all $z \in \mathbb{C}$, then $P(0) = 0$.*

Proof. *(ii) implies (i)* Suppose P satisfies the hypotheses of (i). Then $P_1(z) = P(z + \alpha)$ satisfies the hypotheses of (ii) and $P_1(0) = 0$ implies $P(\alpha) = 0$.

(iii) implies (ii) Suppose P satisfies the hypotheses of (ii). If $P(0) \neq 0$, then if we set $P_1(z) = P(z)/P(0)$, we see that P_1 satisfies the hypotheses of (iii) so $P_1(0) = 0$ and (contrary to our assumption) $P(0) = 0$.

(iv) implies (iii) Suppose P satisfies the hypotheses of (iii). Set $a = P(0)$ and consider the polynomial U defined by $U(z) = P(z) - a$. We have $U(0) = 0$ and so, by Exercise 10.1.9, $U(z) = z^m Q(z)$ with m an integer, $m \geq 1$ and Q a polynomial with $Q(0) \neq 0$. We have $P(z) = a + z^m Q(z)$.

(v) implies (iv) Suppose P satisfies the hypotheses of (iv). Since $Q(0) \neq 0$, Lemma 10.2.2 tells us that there exists a $w \in \mathbb{C}$ with

$$-Q(0)w^m = |Q(0)|.$$

If we set $P_1(z) = P(zw)$, then

$$P_1(z) = a + z^m Q_1(z),$$

with $Q_1(0) = w^m Q(0) = -|Q(0)|$. Thus P_1 satisfies the hypotheses of (v). Since $P_1(0) = 0$ implies $P(0) = 0$, we are done.

(vi) implies (v) Suppose P satisfies the hypotheses of (v). Then, setting $b = Q(0)$ and $U(z) = Q(z) - b$, we have $U(0) = 0$ so $U(z) = zV(z)$ with V a polynomial. Thus P satisfies the hypotheses of (vi).

(i) implies (vi) Trivial. ∎

We can now prove Argand's lemma.

Proof of Theorem 10.2.4. By Lemma 10.2.5, it suffices to show that, if P is a polynomial of the form $P(z) = a - bz^m + z^{m+1} V(z)$ with a, b real, $a \geq 0$, $b > 0$, m an integer, $m \geq 1$ and V a polynomial such that $|P(0)| \leq |P(z)|$ for all $z \in \mathbb{C}$, then $P(0) = 0$, that is to say $a = 0$.

Suppose $a > 0$. Then, since $b(1/n)^m \to 0$ as $n \to \infty$, there exists an N_1 such that $a - b(1/n)^m > 0$ for all $n \geq N_1$ We also observe that

$$\frac{1}{n}V(1/n) \to 0 \times V(0) = 0,$$

so there exists an $N \geq N_1$ such that

$$\frac{b}{2} \geq \left|\frac{1}{n}V(1/n)\right|$$

for all $n \geq N$. If $n \geq N$ we have

$$\begin{aligned}
|P(1/n)| &= |a - b(1/n)^m + (1/n)^{m+1}V(1/n)| \\
&\leq |a - b(1/n)^m| + |(1/n)^{m+1}V(1/n)| \\
&= \left(a - b(1/n)^m\right) + (1/n)^m|(1/n)V(1/n)| \\
&\leq \left(a - b(1/n)^m\right) + (1/n)^m \times (b/2) \\
&= a - (b/2)(1/n)^m < a = P(0).
\end{aligned}$$

Thus, by reductio ad absurdum, we must have $a = 0$ and we are done. ∎

Exercise 10.2.6 *We work in $\mathbb{C}$. Let us write $\mathcal{P}_0$ for the set of polynomials of degree 0 (that is to say the constant non-zero polynomials) and define $\mathcal{P}_n$ inductively by taking $\mathcal{P}_{m+1}$ to be the collection of functions $f : \mathbb{C} \to \mathbb{C}$ given by*

$$f(z) = (z - a)P(z),$$

where $a \in \mathbb{C}$ and $P \in \mathcal{P}_m$.

Show that if $n \geq 1$, then $\mathcal{P}_n$ is the set of polynomials of degree n. (In other words, every polynomial over $\mathbb{C}$ factorises completely into linear factors.)

Exercise 10.2.7 *We work in $\mathbb{C}$.*

(i) *Show that $(z - \alpha)(z - \alpha^*) = z^2 + az + b$, where a and b are real.*

(ii) *Give an inductive definition of what it means for a polynomial of degree n to have real coefficients. (Look at Definition 10.3.1 if you need a hint.)*

(iii) *Show that any polynomial P can be written as $P(z) = P_1(z) + iP_2(z)$ where P_1 and P_2 are polynomials with real coefficients. Deduce that P has real coefficients if and only if $P(x)$ is real whenever x is real.*

(iv) *Suppose P is a polynomial with real coefficients. Show, by induction, that $P(\alpha)^* = P(\alpha^*)$. and deduce that, if $P(\alpha) = 0$, then $P(\alpha^*) = 0$.*

(v) *Suppose P is a polynomial with real coefficients of degree $n \geq 1$. Show that at least one of the following statements must be true.*

(a) *There exists a real β such that $P(z) = (z - \beta)Q(z)$, where Q is a polynomial of degree $n - 1$ with real coefficients.*

(b) *We have* $n \geq 2$ *and there exists an* α *such that*
$$P(z) = (z - \alpha)(z - \alpha^*)V(z),$$
where V *is a polynomial of degree* $n - 2$ *with real coefficients.*

(vi) Deduce that any polynomial of degree $n \geq 1$ *with real coefficients can be written as the product of linear and quadratic polynomials with real coefficients.*

One modern use of the fundamental theorem of algebra in the form given to it in Exercise 10.2.6 occurs in linear algebra, where we use the fact that the characteristic polynomial factorises completely over $\mathbb{C}$. We use the fundamental theorem of algebra in an essential way in Appendix D (see the proof of Theorem D.2).

Note Argand did not prove a result corresponding to Theorem 10.2.3, but assumed that a minimum must exist. From the modern point of view, what he showed was that the very implausible statement 'every polynomial has a root' followed from the very plausible statement that 'the modulus of a polynomial attains a minimum value'.

In the absence of clear foundations for analysis, this is the best that can be done. During his lifetime, Gauss gave four different proofs of the fundamental theorem based on four different (and very fruitful) ideas, but, again, from the modern point of view, they can only show that an implausible statement follows from a plausible statement.

Argand's idea[4] generalises in several directions, and Dirichlet and Riemann used the idea of 'inspecting the minimum' to obtain remarkable results. However, Weierstrass gave examples to show that, in analogous situations to those considered by Dirichlet and Riemann, there may be no minimum.

Fortunately, as we have seen in Theorem 9.3.5 and elsewhere, the Cauchy program completes Argand's proof by giving conditions under which we may be sure that a continuous function has a minimum.[5]

10.3 Liouville Numbers

In this section we will be looking at the real numbers. I have implied that we cannot obtain the real numbers by taking the rational numbers and adjoining roots of appropriate polynomials. This rather vague statement is made precise by a theorem of Liouville.

---

[4] There is no direct connection. Argand's work on the fundamental theorem was not noticed by the general mathematical world.

[5] If the reader has met the standard proof of Rolle's theorem, she will recall that it too depends on 'inspecting a minimum or maximum'.

The limited number of tools available to us means that we require a further definition (compare Definition 10.1.1).

Definition 10.3.1 *We say that* $P : \mathbb{R} \to \mathbb{R}$ *is a polynomial of degree at most* 0 *with integer coefficients if* P *is a constant function with value an integer. If* $n \geq 1$, *we say that* $P : \mathbb{R} \to \mathbb{R}$ *is a polynomial of degree at most n with integer coefficients if* $P(u) = ax^n + Q(x)$ *for all* $x \in \mathbb{R}$, *where a is an integer and* Q *a polynomial with integer coefficients of degree at most* $n - 1$.

Theorem 10.3.2 (Liouville's theorem) *There exists an* $\alpha \in \mathbb{R}$ *such that* α *is not the root of a polynomial with integer coefficients (that is to say,* $P(\alpha) \neq 0$ *whenever* P *is a non-zero polynomial with integer coefficients).*

Exercise 10.3.3 *We define the collection of polynomials with rational coefficients by replacing the word 'integer' by 'rational' in Definition 10.3.1.*

(i) *Show, by induction on the degree of P, that, if P is a polynomial with rational coefficients, we can find an integer* $N \geq 1$ *such that* $U(x) = NP(x)$ *defines a polynomial with integer coefficients.*

(ii) *Deduce that we can replace the words 'integer coefficients' by 'rational coefficients' in Liouville's theorem.*

The first proof of Liouville's theorem that most readers will meet at university is due to Cantor and follows the lines of Theorem 8.2.8 by showing that the collection of real roots of polynomials with integer coefficients can be enumerated and therefore cannot be the whole of $\mathbb{R}$.

Our proof will follow the original proof of Liouville. The key observation is the following.

Theorem 10.3.4 *If P is a non-zero polynomial with integer coefficients and degree at most n and* α *is real but not rational with* $P(\alpha) = 0$, *then we can find an* $A > 0$, *depending on P and* α, *such that*

$$\left| \alpha - \frac{p}{q} \right| > \frac{A}{q^n}$$

whenever p and q are integers with $q \geq 1$.

We say that the non-rational roots of a polynomial are *badly approximable* by rationals.

We need the following simple results.

Exercise 10.3.5

(i) *Let n be an integer with $n \geq 0$. Use induction to show that, if P is a polynomial with integer coefficients and degree at most n and p and q are integers with $q \geq 1$, then $q^n P(p/q)$ is a integer.*

(ii) *Give an inductive proof that, if P is a polynomial and $R > 0$, there exists a $K > 0$ (depending on P and R) such that $|P(x)| \leq K$ whenever $|x| \leq R$. Obtain the same result by using earlier results on continuous functions.*

Proof of Theorem 10.3.4. Set $R = |\alpha| + 1$. Since P has at most n roots (by Theorem 10.1.7), we can find an integer $q_0 \geq 1$ such that, if $|\beta - \alpha| < 2/q_0$ and $P(\beta) = 0$, then $\beta = \alpha$.

Since $P(\alpha) = 0$, we know from our results on factorisation (Theorem 10.1.5) that there exists a polynomial Q such that

$$P(x) = P(x) - P(\alpha) = (x - \alpha)Q(x).$$

By Exercise 10.3.5 (ii), we can find $K > 1$ such that $|Q(x)| \leq K$ for all $|x| \leq R$. Now suppose p and q are integers with $q \geq q_0$. If $|p/q| > R$ then, by our choice of R,

$$\left| \frac{p}{q} - \alpha \right| \geq 1 \geq \frac{1}{q^n}.$$

If $|p/q| \leq R$, we have

$$|P(p/q)| = \left| \left(\frac{p}{q} - \alpha \right) Q(p/q) \right| = \left| \frac{p}{q} - \alpha \right| |Q(p/q)| \leq K \left| \frac{p}{q} - \alpha \right|. \quad \bigstar$$

Since $q \geq q_0$, either

$$\left| \frac{p}{q} - \alpha \right| \geq \frac{2}{q_0},$$

so that

$$\left| \frac{p}{q} - \alpha \right| \geq \frac{1}{q^n},$$

automatically, or $P(p/q) \neq 0$. In the second case, Exercise 10.3.5 (i) tells us that $q^n P(p/q)$ is a non-zero integer and so, since 1 is the least strictly positive integer, $|q^n P(p/q)| \geq 1$. The inequality $\bigstar$ now gives

$$K \left| \frac{p}{q} - \alpha \right| \geq \frac{1}{q^n}.$$

We have shown that the inequality just stated holds for all $q \geq q_0$. Since α is irrational, we can find an A with $K^{-1} > A > 0$ such that

$$\left| \frac{p}{q} - \alpha \right| > A$$

for all integers p and each integer q with $q_0 > q \geq 1$. We now have

$$\left| \alpha - \frac{p}{q} \right| > \frac{A}{q^n}$$

whenever p and q are integers with $q \geq 1$, as required. ∎

The approximation theorem just stated allows us to write down a number which is not the root of a polynomial with integer coefficients, and thus prove Liouville's theorem.

Lemma 10.3.6 *If $a_0 = 0$ and $a_r = a_{r-1} + 10^{-r!}$, then $a_r \to \alpha$ as $r \to \infty$, where α is not the root of a polynomial with integer coefficients.*

Exercise 10.3.7 *Write down the number α just defined in decimal notation to 30 places of decimals.*

Proof of Lemma 10.3.6. Observe that, if $k > r$, then $10^{-k!} \leq 10^{-(r+1)!} \times 10^{-k+r+1}$ and so (by induction or summing a geometric series)

$$a_r \leq a_s \leq a_r + \frac{10}{9}(1 - 10^{r-s})10^{-(r+1)!} \leq a_r + 2 \times 10^{-(r+1)!}$$

for all $s \geq r$. Thus the a_s form an increasing sequence bounded above and so $a_s \to \alpha$ for some $\alpha \in \mathbb{R}$. Further, we have

$$a_r \leq \alpha \leq a_r + 2 \times 10^{-(r+1)!}.$$

A simple induction shows that $10^{r!}a_r$ is an integer. Thus, if we write $q_r = 10^{r!}$ and $p_r = 10^{r!}a_r$, we have p_r and q_r integers with $q_r \geq 1$ and

$$\left| \alpha - \frac{p_r}{q_r} \right| = |\alpha - a_r| \leq \frac{2}{10^{(r+1)!}} = \frac{2}{q_r^{r+1}}.$$

Thus, if we choose any integer n with $n \geq 1$ and any real number A with $A > 0$, we have

$$\left| \alpha - \frac{p_r}{q_r} \right| < \frac{A}{q_r^n}$$

for r sufficiently large. Since A was arbitrary, Theorem 10.3.4, tells us that α cannot be the root of a polynomial of degree n or less with integer coefficients.

Since n was arbitrary, α cannot be the root of any polynomial with integer coefficients. ∎

Exercise 10.3.8 *Let $e(j)$ take the value 1 or 2 $[j \geq 1]$ (for example, we could take $e(1) = 1$, $e(2) = 2$, $e(3) = 2$, $e(4) = 1$ and so on). If $x_0 = 0$ and $x_n = x_{n-1} + e(n)10^{-n!}$, show that $x_n \to x$ as $n \to \infty$ where x is not the root of a polynomial with integer coefficients.*

It is natural to ask for an example 'not constructed for the purpose'. Hermite showed the number e is not the root of a polynomial with integer coefficients (again by showing that it is too well approximable) and Lindemann extended Hermite's argument to show that this is also true for π.

10.4 A Non-Archimedean Ordered Field

All the ordered fields that we have looked at so far can be considered as sub-field of $\mathbb{R}$ and so are automatically Archimedean. In this final section of the chapter, we use polynomials to construct a 'big ordered field' in which the axiom of Archimedes fails.

Since we are only doing our construction to show that such an object exists and not because we wish to use the object, I suggest that the reader does not work too hard, but merely seeks to understand the, by now, familiar kind of construction we use.

Definition 10.4.1 *Let $(\mathcal{A}, +, \times)$ be a set together with operations $+$ and times $\times$ such that, whenever P, $Q \in \mathcal{A}$ we have $P + Q$, $P \times Q \in \mathcal{A}$. We say that $(\mathcal{A}, +, \times)$ is an* integral domain *if, whenever P, Q, $R \in \mathcal{A}$, the following results hold.*

(i) $P + Q = Q + P$. (Commutative law of addition.)
(ii) $P + (Q + R) = (P + Q) + R$. (Associative law of addition)
(iii) There exists a $0 \in \mathcal{A}$ such that $0 + P = P$ for all P. (Additive zero)
(iv) For each P, we can find $-P$ such that $P + (-P) = 0$. (Additive inverse)
(v) $P \times Q = Q \times P$. (Commutative law of multiplication)
(vi) $P \times (Q \times R) = (P \times Q) \times R$. (Associative law of multiplication)
(vii) There exists a $1 \in \mathcal{A}$ such that $1 \times P = P$ for all P. (Multiplicative unit)
(viii) If $P \neq 0$ and $P \times Q = 0$ then $Q = 0$. (Multiplicative cancellation law)
(ix) $P \times (Q + R) = (P \times Q) + (P \times R)$. (Distributive law)

We also demand $0 \neq 1$.

If the reader compares the laws for an integral domain with the laws for a field, she will see that they are the same, except that the existence of a multiplicative inverse is replaced by the weaker multiplicative cancellation law.

Exercise 10.4.2 *(The reader should recognise this as standard.) State and prove theorems to the effect that the zero and unit of an integral domain are unique. Show also that the additive inverse is unique.*

Exercise 10.4.3 *Let $\mathcal{R}$ be an integral domain and let P, Q, $R \in \mathcal{R}$.*

(i) Show that $0 \times R = 0$.
(ii) Show that, if $P \neq 0$ and $P \times Q = P \times R$, then $Q = R$.

We now consider the collection of polynomials $\mathcal{P}$ over a field $\mathbb{F}$, where $\mathbb{F}$ is $\mathbb{R}$, $\mathbb{Q}$ or $\mathbb{C}$. Notice that, if P, $Q \in \mathcal{P}$, then, by Exercise 10.1.3, we have $P + Q$, $P \times Q$, $-P \in \mathcal{P}$.

Exercise 10.4.4 *Let $P(u) = u$. Show that there does not exist a $Q \in \mathcal{P}$ such that $P \times Q = 1$. (Thus $\mathcal{P}$ is not a field.)*

Theorem 10.4.5 *$(\mathcal{P}, +, \times)$ is an integral domain.*

Proof. With the exception of the multiplicative cancellation law, all the proofs are immediate. We illustrate with (i) and leave the rest (apart from (viii)) to the reader.
Proof of (i) Since $\mathbb{F}$ is a field, we have

$$(P + Q)(u) = P(u) + Q(u) = Q(u) + P(u) = (Q + P)(u)$$

for all $u \in \mathbb{F}$ and so $P + Q = Q + P$.
Proof of (viii) Suppose that P and Q are non-zero polynomials. Then P is a polynomial of degree $n \geq 0$ with leading coefficient $a \neq 0$ and Q is a polynomial of degree $m \geq 0$ with leading coefficient $b \neq 0$. By Lemma 10.1.3, $P \times Q$ is a polynomial of degree $n + m$ with leading coefficient $ab \neq 0$. Thus, by Exercise 10.1.10 (an easy consequence of Theorem 10.1.7), there exists a $t \in \mathbb{R}$ such that $(P \times Q)(t) \neq 0$ and so $P \times Q \neq 0$. ∎

Exercise 10.4.6 *Check (possibly in your head) that the remaining rules for integral domains are satisfied by $(\mathcal{P}, +, \times)$.*

We are also interested in order.

Definition 10.4.7 *We say that an integral domain*$(\mathcal{A}, +, \times)$ *together with a relation* $>$ *is an* ordered integral domain *if the following rules hold for* $P, Q, R \in \mathcal{A}$.

(x) *If* $P > Q$ *and* $Q > R$, *then* $P > R$. *(Transitivity of order)*
(xi) *Exactly one of the following conditions holds:* $P > Q$ *or* $Q > P$ *or* $P = Q$. *(Trichotomy)*
(xii) *If* $P > Q$, *then* $P + R > Q + R$. *(Order and addition)*
(xiii) *If* $P > Q$ *and* $R > 0$, *then* $P \times R > Q \times R$. *(Order and multiplication)*

Notice that $(\mathbb{Z}, +, \times, >)$ is an ordered integral domain. If we now specialise to the case $\mathbb{F} = \mathbb{R}$, the collection $\mathcal{P} = \mathcal{P}_{\mathbb{R}}$ turns out not only to be an integral domain, but an ordered integral domain for the appropriate definition of order.

Definition 10.4.8 *Let* $P \in \mathcal{P}_{\mathbb{R}}$. *We say that* $P \succ 0$ *if the leading coefficient* a *of* P *satisfies* $a > 0$.
If $P, Q \in \mathcal{P}_{\mathbb{R}}$ *and* $P - Q \succ 0$, *we say that* $P \succ Q$.

Exercise 10.4.9 *Give an example of a polynomial* P *such that* $P \succ 0$, *but* $P(0) < 0$.

Theorem 10.4.10 $(\mathcal{P}_{\mathbb{R}}, +, \times, \succ)$ *is an ordered integral domain.*

Proof. (x) If $P \succ Q$ and $Q \succ R$, then $P - Q$ has leading coefficient $a > 0$ and $Q - R$ has leading coefficient $b > 0$. Depending on the degree of $P - Q$ and $Q - R$, this implies that the leading coefficient of

$$P - R = (P - Q) + (Q - R)$$

is a, b or $a + b$. Since a, b, $a + b > 0$, it follows that $P - R \succ 0$ and so $P \succ R$.

(xi) Either $P - Q = 0$, so $P = Q$ and $P \not\succ Q$, $Q \not\succ P$ or $P - Q$ has leading coefficient $c \neq 0$. If $c > 0$, then $P \succ Q$ and $Q \not\succ P$, $P \neq Q$. If $c < 0$, then $Q \succ P$ and $P \not\succ Q$, $P \neq Q$.

(xii) If $P \succ Q$, then

$$(P + R) - (Q + R) = P - Q \succ 0,$$

so $P + R \succ Q + R$.

(xiii) If $P \succ Q$ and $R \succ 0$, then $P - Q$ has leading coefficient $a > 0$ and R has leading coefficient $b > 0$. Thus

$$P \times R - Q \times R = (P - Q) \times R$$

has leading coefficient $ab > 0$ and $P \times R \succ Q \times R$. ∎

The reader who recalls the construction $\mathbb{Q}$ will see a clear path from integral domains to fields. We sketch this path leaving the reader to perform any verifications she wishes.

Exercise 10.4.11 *Let* $(\mathcal{A}, +, \times)$ *be an integral domain and write* $\mathcal{B}$ *for the set of ordered pairs* (P, Q) *with* P, $Q \in \mathcal{A}$ *and* $Q \neq 0$. *We define a relation* $\sim$ *on* $\mathcal{B}$ *by the condition* $(P, Q) \sim (R, S)$ *if and only if* $P \times S = R \times Q$. *Show that* $\sim$ *is an equivalence relation.*
[See Lemma 2.2.7 if you need a hint.]

Exercise 10.4.12 *Consider* $(\mathbb{Z}_6, +, \times)$, *the system of integers modulo 6. Write* $\mathcal{B}$ *for the set of ordered pairs* $([u], [v])$ *with* $[u]$, $[v] \in \mathbb{Z}_6$ *and* $[v] \neq [0]$. *We define a relation* $\sim$ *on* $\mathcal{B}$ *by the condition* $([a], [b]) \sim ([c], [d])$ *if and only if* $[a] \times [d] = [b] \times [c]$. *Show that*

$$([2], [2]) \sim ([3], [3]), \ ([3], [3]) \sim ([4], [2]), \ but \ ([2], [2]) \nsim ([4], [2]).$$

Why does the result of Exercise 10.4.11 fail?

Exercise 10.4.13 *We continue with the chain of thought of Exercise 10.4.11. Let* (P_i, Q_i), $(R_i, U_i) \in \mathcal{B}$ *and suppose that* $(P_1, Q_1) \sim (P_2, Q_2)$ *and* $(R_1, U_1) \sim (R_2, U_2)$.

(i) Show that

$$(P_1 \times R_1) \times (Q_2 \times U_2) = (P_2 \times R_2) \times (Q_1 \times U_1)$$

and deduce that $(P_1 \times R_1, Q_1 \times U_1) \sim (P_2 \times R_2, Q_2 \times U_2)$.
(ii) Conclude that we can define multiplication on $\mathcal{B}/\sim$ *by*

$$[(P, Q)] \times [(R, U)] = [(P \times R, Q \times U)].$$

(iii) Show that

$$\big((P_1 \times U_1) + (Q_1 \times R_1)\big) \times (Q_2 \times U_2) = \big((P_2 \times U_2) + (Q_2 \times R_2)\big) \times (Q_1 \times U_1).$$

(iv) Conclude that we can define addition on $\mathcal{B}/\sim$ *by*

$$[(P, Q)] + [(R, U)] = [((P \times U) + (Q \times R), Q \times U)].$$

[See Lemma 2.2.11 if you need a hint.]

Theorem 10.4.14 *With the notation of the previous exercises,* $(\mathcal{B}/\sim, +, \times)$ *is a field with zero* $[(0, 1)]$; *unit* $[(1, 1)]$; *additive inverse of* $[(P, Q)]$, *the element* $[(-P, Q)]$ *and multiplicative inverse of a non-zero element* $[(P, Q)]$, *the element* $[(Q, P)]$.

Proof. Left to the reader. ■

Exercise 10.4.15 *Carry out as much of the proof of Theorem 10.4.14 as you feel is necessary.*

If we start with an ordered integral domain, we end with an ordered field.

Exercise 10.4.16 *Let* $(\mathcal{A}, +, \times, >)$ *be an ordered integral domain. We use the notation already established.*

(i) Show that, if $P > 0$, *then* $0 > -P$ *and, if* $0 > P$, *then* $-P > 0$.

(ii) Show that, if $0 > P, Q$, *then* $P \times Q > 0$, *but, if* $P > 0 > Q$, *then* $0 > P \times Q$.

(iii) Show that $(P, Q) \sim (0, 1)$ *if and only if* $P = 0$.

(iv) Show that, if $(P, Q) \sim (R, U)$, *then, if* $P \times Q > 0$, *it follows that* $R \times U > 0$.

(v) Conclude that we may define $[(P, Q)] > [(0, 1)]$ *by the condition* $P \times Q > 0$.

Theorem 10.4.17 *We use the notation of Theorem 10.4.14. Suppose that we work with an ordered integral domain* $(\mathcal{A}, +, \times, >)$. *Show that, if we extend the definition of Exercise 10.4.16 by writing*

$$[(P, Q)] > [(R, U)] \text{ if } [(P, Q)] - [(R, U)] > [(0, 1)],$$

then $(\mathcal{B}/\sim, +, \times, >)$ *is an ordered field.*

Proof. Left to the reader. ■

Exercise 10.4.18 *Carry out as much of the proof of Theorem 10.4.17 as you feel is necessary.*

Theorem 10.4.19 *If we perform the construction outlined in this section with the ordered integral domain* $\mathcal{A} = \mathcal{P}_{\mathbb{R}}$ *formed by the real polynomials, the resulting ordered field* $\mathcal{F} = \mathcal{B}/\sim$ *does not satisfy the axiom of Archimedes.*

Proof. Observe that $\theta : \mathbb{Q} \to \mathcal{F}$ given by $\theta(m/n) = [(v_m, v_n)]$ (where v_k is the constant polynomial $v_k(x) = k$) for $m, n \in \mathbb{Z}$, $n \neq 0$ gives the unique isomorphism from $\mathbb{Q}$ to a sub-field of $\mathcal{F}$. If we let $U(x) = x$ for all $x \in \mathbb{R}$ then

$$\theta(1/n) = [(v_1, v_n)] > [(1, U)] > 0$$

for all $n \in \mathbb{Z}$, $n \geq 1$. Thus $\theta(1/n) \nrightarrow 0$ as $n \to \infty$ and the axiom of Archimedes fails. ■

Remark The reader may ask why we do not simply write $P(x)/Q(x)$ in place of $[(P, Q)]$. She should consider the problem of assigning meaning to $1/x$ or to x^2/x^4 when $x = 0$.

Exercise 10.4.20 *This fairly long exercise gives what might be called an algebraist's construction of* $\mathbb{C}$. *Throughout,* $\mathbb{F}$ *is one of* $\mathbb{Q}$, $\mathbb{R}$ *or* $\mathbb{C}$. *We suppose* $v \in \mathbb{F}$ *and work with the collection* $\mathcal{P} = \mathcal{P}_{\mathbb{F}}$ *of polynomials over* $\mathbb{F}$.

(i) *Show, by induction, that any* $P \in \mathcal{P}$ *can be written as*

$$P(t) = Q(t) \times (t^2 - v) + (at + b)$$

for some $Q \in \mathcal{P}$ *and* a, $b \in \mathbb{F}$.

(ii) *Show, by using Exercise 10.1.10, that* a *and* b *are uniquely defined in (i) once* P *is given.*

(iii) *Let us write* $P_1 \sim_v P_2$ *if*

$$P_1(t) - P_2(t) = Q(t) \times (t^2 - v)$$

with $Q \in \mathcal{P}$. *Show that* $\sim_v$ *is an equivalence relation. We write* $[P]$ *for the equivalence class of* P.

(iv) *Show that, if* $P_1 \sim_v P_2$ *and* $R_1 \sim_v R_2$, *then* $P_1 + R_1 \sim_v P_2 + R_2$ *and* $P_1 \times R_1 \sim_v P_2 \times R_2$. *Conclude that we may make the unambiguous definitions*

$$[P] + [R] = [P + R] \text{ and } [P] \times [R] = [P \times R].$$

(v) *Show that, with this addition and multiplication,* $\mathcal{P}/\sim_v$ *obeys all the rules for an integral domain (see Definition 10.4.1) with the possible exception of (viii). (Note that this is as easy as it looks.)*

(vi) *Suppose that* $\mathbb{F} = \mathbb{R}$ *and* $v = 1$. *By observing that, if* $P(u) = u - 1$ *and* $Q(u) = u + 1$, *then* $[P] \times [Q] = [0]$, *show that* $\mathcal{P}/\sim_v$ *is not an integral domain. Obtain similar results when* $\mathbb{F} = \mathbb{R}$ *and* $v \geq 0$ *and when* $\mathbb{F} = \mathbb{C}$ *and* $v = -1$.

(vii) *Suppose that* $\mathbb{F} = \mathbb{R}$ *and* $v = -1$. *Show that* $\mathcal{P}/\sim_v$ *is a field by observing that, if* $P(u) = au + b$ *with* a *and* b *not both zero and we set*

$$Q(u) = \frac{-a}{a^2 + b^2} u + \frac{b}{a^2 + b^2},$$

then $[P] \times [Q] = [1]$. *Obtain a similar result when* $\mathbb{F} = \mathbb{Q}$ *and* $v = 2$.

(viii) *Continuing with the ideas of (vi), show that, if* $\mathbb{F} = \mathbb{R}$ *and* $v = -1$, *the function* $f : \mathbb{C}/\sim_v \to \mathcal{P}/\sim_v$ *defined by* $f(b + ai) = [Q]$ *with*

$Q(t) = at + b$ *(where a, b $\in \mathbb{R}$) is a bijection which preserves addition and multiplication (that is to say a field isomorphism).*

[*Remark. The construction of $\mathbb{C}$ using equivalence classes has the advantage of lending itself to many generalisations. We can take any field in place of the three we considered and any polynomial in place of $t^2 + v$ and see what happens. The disadvantage is that we lose the geometric interpretation on which complex analysis relies. For the analyst this is a very serious loss.*]

11
Can We Go Further?

11.1 The Quaternions

Once it is realised that the complex numbers can actually be constructed from the reals and that they form a useful tool for geometry and physics, it is natural to seek further generalisations. The complex numbers give an interesting algebraic structure associated with the two-dimensional plane $\mathbb{R}^2$, but we live in the three-dimensional[1] space $\mathbb{R}^3$. Can we repeat the trick by looking not at $x + yi$, but at $x + yi + wj$?

Among those who pondered this question was the great Irish mathematician Hamilton. He seems to have thought about it, on and off, for ten years leading up to a climax described in the following letter to his elder son.

Every morning in the early part of [October 1843], on my coming down to breakfast, your (then) little brother William Edwin, and yourself, used to ask me, 'Well, Papa, can you *multiply* triplets ?' Whereto I was always obliged to reply, with a sad shake of the head 'No, I can only add and subtract them'.

But on the 16th day of the same month which happened to be Monday, and a Council day of the Royal Irish Academy, I was walking in to attend and preside, and your mother was walking with me, along the Royal Canal, to which she had perhaps driven and although she talked with me now and then, yet an under-current of thought was going on in my mind, which gave at last a result, whereof it is not too much to say that I felt at once the importance. An electric circuit seemed to close; and a spark flashed forth, the herald (as I foresaw, immediately) of many long years to come of definitely directed thought and work, by myself if spared, and at all events on the part of others, if I should even be allowed to live long enough distinctly to communicate the discovery. Nor could I resist the

---

[1] Potential readers of this chapter are unlikely not to recognise $\mathbb{R}^n$ as the collection of ordered n-tuples of reals.

impulse – unphilosophical as it may have been – to cut with a knife on a stone of [Broome] Bridge, the fundamental formula with the symbols, i, j, k namely

$$i^2 = j^2 = k^2 = ijk = -1.$$

In Volume II of Grave's Life of Hamilton *[11]*

Hamilton's new system, which he called the quaternions (often denoted by $\mathbb{H}$), involved not 'triplets' but 'quadruplets'

$$\mathbf{x} = x_0 + x_1 i + x_2 j + x_3 k$$

with addition, as might be expected, given by

$$\mathbf{x} + \mathbf{y} = (x_0 + y_0) + (x_1 + y_1)i + (x_2 + y_2)j + (x_3 + y_3)k.$$

However, multiplication used the rules

$$i^2 = j^2 = k^2 = -1, \ ij = k, \ ji = -k, \ jk = i, \ kj = -i, \ ki = j, \ ik = -j,$$

so that

$$\mathbf{x} \otimes \mathbf{y} = (x_0 y_0 - x_1 y_1 - x_2 y_2 - x_3 y_3) + (x_0 y_1 + x_1 y_0 + x_2 y_3 - x_3 y_2)i$$
$$+ (x_0 y_2 + x_2 y_0 + x_3 y_1 - x_1 y_3)j + (x_0 y_3 + x_3 y_0 + x_1 y_2 - x_2 y_1)k.$$

Exercise 11.1.1

(i) Check that the last sentence makes sense (at least informally).
(ii) Check (again informally) that we can derive the rules

$$i^2 = j^2 = k^2 = -1, \ ij = k, \ ji = -k, \ jk = i, \ kj = -i, \ ki = j, \ ik = -j$$

from the rules

$$i^2 = j^2 = k^2 = ijk = -1$$

and vice versa.

We may remember the usefulness of the conjugacy operation for $\mathbb{C}$ (see Definition 9.2.3) and define conjugacy for quaternions by

$$\mathbf{x}^\star = x_0 - x_1 i - x_2 j - x_3 k.$$

Part (i) of the next exercises formalises our construction for the quaternions and part (ii) gives an alternative construction.

Exercise 11.1.2 *Consider the systems* $(\mathbb{H}_1, +, \otimes, ^\star)$ *and* $(\mathbb{H}_2, +, \otimes, ^\star)$ *defined as follows:*

(1) $\mathbb{H}_1$ *consists of ordered quadruples* $\mathbf{x} = (x_0, x_1, x_2, x_3)$ *of real numbers and we use the rules*

$$\mathbf{x} + \mathbf{y} = (x_0 + y_0, x_1 + y_1, x_2 + y_2, x_3 + y_3)$$

$$\mathbf{x} \otimes \mathbf{y} = (x_0 y_0 - x_1 y_1 - x_2 y_2 - x_3 y_3, x_0 y_1 + x_1 y_0 + x_2 y_3 - x_3 y_2,$$

$$x_0 y_2 + x_2 y_0 + x_3 y_1 - x_1 y_3, x_0 y_3 + x_3 y_0 + x_1 y_2 - x_2 y_1)$$

$$\mathbf{x}^\star = (x_0, -x_1, -x_2, -x_3).$$

(2) $\mathbb{H}_2$ *consists of ordered pairs* $\mathbf{z} = (z_1, z_2)$ *of complex numbers and we use the rules*

$$\mathbf{z} + \mathbf{w} = (z_1 + w_1, z_2 + w_2)$$

$$\mathbf{z} \otimes \mathbf{w} = (z_1 w_1 - z_2 w_2^*, z_1 w_2 + z_2 w_1^*)$$

$$\mathbf{z}^\star = (z_1^*, -z_2).$$

Show that the function $f : \mathbb{H}_1 \to \mathbb{H}_2$ *given by*

$$f\big((x_0, x_1, x_2, x_3)\big) = (x_0 + ix_1, x_2 + ix_3)$$

is a bijection that preserves $+$, $\otimes$ *and* *, *that is to say, an isomorphism.*

We shall use whichever (isomorphic) model of the quaternions seems most appropriate. Notice that our version (2) mimics our earlier construction of $\mathbb{C}$ from $\mathbb{R}$.

Remarkably, Hamilton's insight gives us a system with all the properties of a field *except* that multiplication ceases to be commutative.

Definition 11.1.3 *A skew-field* $(\mathbb{A}, +, \otimes)$ *is a set* $\mathbb{A}$ *together with two operations* $+$ *and* $\otimes$ *(with* $\mathbf{a} + \mathbf{b} \in \mathbb{A}$, $\mathbf{a} \otimes \mathbf{b} \in \mathbb{A}$ *whenever* $\mathbf{a}$, $\mathbf{b} \in \mathbb{A}$) *having the following properties.*

(i) $\mathbf{a} + \mathbf{b} = \mathbf{b} + \mathbf{a}$. *(Commutative law of addition)*

(ii) $\mathbf{a} + (\mathbf{b} + \mathbf{c}) = (\mathbf{a} + \mathbf{b}) + \mathbf{c}$. *(Associative law of addition)*

(iii) *There exists an element* $0 \in \mathbb{A}$ *such that* $0 + \mathbf{a} = \mathbf{a}$. *(Additive zero)*

(iv) *For each* $\mathbf{a} \in \mathbb{A}$, *there exists an element* $-\mathbf{a} \in \mathbb{A}$ *such that* $\mathbf{a} + (-\mathbf{a}) = 0$. *(Additive inverse)*

(v) $\mathbf{a} \otimes (\mathbf{b} \otimes \mathbf{c}) = (\mathbf{a} \otimes \mathbf{b}) \otimes \mathbf{c}$. *(Associative law of multiplication)*

(vi) *There exists an element* $1 \in \mathbb{A}$ *such that* $1 \otimes \mathbf{a} = \mathbf{a} \otimes 1 = \mathbf{a}$. *(Multiplicative unit)*

(vii) If $\mathbf{a} \neq 0$, then there exists a $\mathbf{a}^{-1} \in \mathbb{A}$ such that $\mathbf{a} \otimes \mathbf{a}^{-1} = \mathbf{a}^{-1} \otimes \mathbf{a} = 1$.
(Multiplicative inverse)
(viii) We have

$$\mathbf{a} \otimes (\mathbf{b} + \mathbf{c}) = (\mathbf{a} \otimes \mathbf{b}) + (\mathbf{a} \otimes \mathbf{c})$$

and

$$(\mathbf{b} + \mathbf{c}) \otimes \mathbf{a} = (\mathbf{b} \otimes \mathbf{a}) + (\mathbf{c} \otimes \mathbf{a}).$$

(Distributive laws)

We also demand $1 \neq 0$.

Comparing Definition 11.1.3 with Definition 5.1.1, we see that the omission of the commutative law for multiplication has led us to modify the law governing the multiplicative unit (vi), the law governing the multiplicative inverse (vii) and the distributive law (viii).

Exercise 11.1.4 *State results corresponding to Exercise 3.2.11 for a skew-field and prove them.*

Theorem 11.1.5 *The quaternions form a skew-field.*

We split the proof into two steps. We start with some direct verifications.

Exercise 11.1.6 *Show that $(\mathbb{H}, +, \otimes)$ satisfies the conditions of Definition 11.1.3, with the possible exception of the existence of a multiplicative inverse.*

[The calculations are slightly easier if we use the definition of quaternions in terms of ordered pairs of complex numbers.]

Our proof of the existence of a multiplicative inverse will be embedded in the study of quaternion conjugation.

Exercise 11.1.7 *In this exercise, we consider quaternions as ordered quadruples of the reals $\mathbf{x} = (x_0, x_1, x_2, x_3)$ [following representation (1) of Exercise 11.1.2] and write*

$$\|\mathbf{x}\| = \sqrt{x_0^2 + x_1^2 + x_2^2 + x_3^2}$$

(where we take the positive square root).

(i) If $\mathbf{i} = (0, 1, 0, 0)$ and $\mathbf{j} = (0, 0, 1, 0)$, verify that $(\mathbf{i} \otimes \mathbf{j})^\star \neq \mathbf{i}^\star \otimes \mathbf{j}^\star$.
(ii) Show that $(\mathbf{x} \otimes \mathbf{y})^\star = \mathbf{y}^\star \otimes \mathbf{x}^\star$.

(iii) Show that

$$\mathbf{y} \otimes \mathbf{y}^\star = \mathbf{y}^\star \otimes \mathbf{y} = (\|\mathbf{y}\|^2, 0, 0, 0).$$

(iv) If $\mathbf{x} \neq (0, 0, 0, 0)$ *and we write*

$$\mathbf{x}^{-1} = \left(\frac{x_0}{\|\mathbf{x}\|^2}, -\frac{x_1}{\|\mathbf{x}\|^2}, -\frac{x_2}{\|\mathbf{x}\|^2}, -\frac{x_3}{\|\mathbf{x}\|^2} \right),$$

show that

$$\mathbf{x} \otimes \mathbf{x}^{-1} = \mathbf{x}^{-1} \otimes \mathbf{x} = (1, 0, 0, 0).$$

(v) If $\mathbf{x}, \mathbf{y} \neq (0, 0, 0, 0)$, *show that* $(\mathbf{x} \otimes \mathbf{y})^{-1} = \mathbf{y}^{-1} \otimes \mathbf{x}^{-1}$.
(vi) Give an example of $\mathbf{x}, \mathbf{y} \neq (0, 0, 0, 0)$ *such that* $(\mathbf{x} \otimes \mathbf{y})^{-1} \neq \mathbf{x}^{-1} \otimes \mathbf{y}^{-1}$.
(vii) Show that $(\mathbf{x} + \mathbf{y})^* = \mathbf{x}^* + \mathbf{y}^*$ *and* $(\mathbf{x}^*)^\star = \mathbf{x}$.

Part (iv) of the previous exercise completes the proof that the quaternions form a skew-field.

Exercise 11.1.8 *Version (2) of Exercise 11.1.2 looks a bit complicated. In this exercise we see what happens if we try to simplify things by imitating the construction of* $\mathbb{C}$ *from* $\mathbb{R}$ *exactly.*

Suppose that $\mathcal{H}$ *consists of ordered pairs* $\mathbf{z} = (z_1, z_2)$ *of complex numbers and we use the rules*

$$\mathbf{z} + \mathbf{w} = (z_1 + w_1, z_2 + w_2)$$
$$\mathbf{z} \boxtimes \mathbf{w} = (z_1 w_1 - z_2 w_2, z_1 w_2 + z_2 w_1).$$

Explain briefly why (taking $(1, 0)$ *as our multiplicative unit and* $(0, 0)$ *as our additive zero)* $(\mathcal{H}, +, \boxtimes)$ *satisfies all the rules for a field given in Definition 5.1.1 with the exception of the existence of a multiplicative inverse (condition (viii)). Show, however, that*

$$(1, i) \boxtimes (i, 1) = (0, 0)$$

and deduce that (viii) fails.

Exercise 11.1.9 *We continue with the ideas of Exercise 11.1.7.*

(i) By considering $(\mathbf{x} \otimes \mathbf{y}) \otimes (\mathbf{x} \otimes \mathbf{y})^*$, *show that*

$$\|\mathbf{x} \otimes \mathbf{y}\| = \|\mathbf{x}\| \|\mathbf{y}\|.$$

(ii) Deduce that, if n_j *and* m_j *are integers* $[0 \leq j \leq 3]$, *we can find integers* q_j *such that*

$$q_0^2 + q_1^2 + q_2^2 + q_3^2 = (n_0^2 + n_1^2 + n_2^2 + n_3^2) \times (m_0^2 + m_1^2 + m_2^2 + m_3^2).$$

(This result goes back to Euler.)

Exercise 11.1.10 *The lack of a commutative law of multiplication can have unexpected consequences. In this exercise we consider quaternions as ordered quadruplets of the reals* $\mathbf{x} = (x_0, x_1, x_2, x_3)$.

(i) *Write down*

$$\mathbf{x}^2 = \mathbf{x} \otimes \mathbf{x}$$

as an ordered quadruplet.

(ii) *Solve the equation* $\mathbf{x}^2 = (1, 0, 0, 0)$, *showing that there are exactly two solutions.*

(iii) *Solve the equation* $\mathbf{x}^2 = (-1, 0, 0, 0)$, *showing that there are infinitely many solutions.*

[*Some of the strangeness of part (iii) vanishes when we recall that our treatment of quadratics over a field involves equations like*

$$(x + a)(x + b) = x^2 + ax + xb + ab = x^2 + (a + b)x + ab$$

which may cease to be true in the absence of a commutative law of multiplication.]

Exercise 11.1.11 *Speaking very roughly, Exercise 7.2.7 tells us that there is only one way of embedding the rationals in the reals in a natural manner and Exercise 9.3.4 tells us that there is only one way of embedding the reals in the complex numbers. However, this exercise shows that are many ways of embedding the complex numbers in the quaternions.*

Let $\mathbf{u}$ *be a solution of the equation* $\mathbf{x}^2 = (-1, 0, 0, 0)$ *discussed in part (iii) of the previous exercise. If we define* $f : \mathbb{C} \to \mathbb{H}$ *by* $f(a + ib) = a + b\mathbf{u}$ *for a and b real prove that* f *is injective and*

$$f(z_1 + z_2) = f(z_1) + f(z_2), \quad f((z_1 \times z_2)) = f(z_1) \otimes f(z_2), \quad f(z^*) = f(z)^*$$

for all z_1, z_2.
[*Exercise C.6 explores related ideas.*]

11.2 What Happened Next

Hamilton's quaternions created a sensation. Older systems (even the complex numbers) had been around long enough that people thought of them as 'natural systems' which had been *discovered*. Hamilton's new numbers looked very much as if they had been *created*. Laws like commutativity which (if anybody thought about them at all[2]) were seen as embedded in the nature of things, now became properties which might, or might not, be possessed by an operation. Adjectives like associative (invented by Hamilton himself) and distributive had to be added to the mathematician's vocabulary.

However, Hamilton had invented his system for use rather than admiration. Complex numbers were two-dimensional, and the world is three-dimensional. Hamilton had sought 'triplets' to match the three dimensions. His new system consisted of quadruplets. Fortunately, as the reader has probably already noticed, it contains a very natural three-dimensional object. If, instead of looking at

$$\mathbf{x} = x_0 + x_1 i + x_2 j + x_3 k,$$

we consider

$$\mathbf{x} = x_0 + \underline{x}$$

with $\underline{x} = x_1 i + x_2 j + x_3 k$ then $\underline{x}$ corresponds in an obvious way to the point with Cartesian coordinates (x_1, x_2, x_3). Hamilton called x_0 the *scalar part* and $\underline{x}$ the *vector part* of the quaternion $\mathbf{x}$.

Hamilton wrote many papers explaining the geometric and physical uses of quaternions, but failed to convince most practising mathematicians and physicists of the usefulness, as opposed to the ingenuity, of his system. One of those prepared to invest time and energy in mastering the new system was Tait, who helped to extend the appropriate calculus notions to quaternions. (It was only when calculus was developed for the complex numbers by Cauchy and his students that they became an important tool for physicists.) Maxwell was a friend and correspondent of Tait, and the introductory chapter of his revolutionary *Treatise on Electricity and Magnetism* contains the following passage:

> ... for many purposes in physical reasoning apart from calculation, it is desirable to avoid explicitly introducing the Cartesian coordinates, and to fix the mind at once on a point of space instead of its three coordinates, and on the magnitude and direction of a force instead of its three components. This mode of contemplating geometrical and physical quantities is more primitive and more natural than the other, although the ideas connected with it did not receive their full development till Hamilton made the next great step in dealing with space by the invention of his Calculus of Quaternions
>
> ... I am convinced ... that the introduction of the ideas, as distinguished from the operations and methods of Quaternions, will be of great use to us in the study of all parts of our subject, and especially in electrodynamics, where we have to deal with

---

[2] Euclid did. In modern notation, Proposition 17 of Book 7 in [8] states that $n \times m = m \times n$.

a number of physical quantities, the relations of which to each other can be expressed far more simply by a few words of Hamilton, than by the ordinary equations.

Maxwell did not use quaternions for calculation, but presented some of his results and, in particular, the fundamental equations of electromagnetism, in quaternionic form. Physicists like Gibbs and Heaviside carried the program further, but, rather than use the entire quaternion $\mathbf{x}$, they simply used the vector part $\underline{x}$.

If we write $\underline{x}$ as a triple

$$\underline{x} = (x_1, x_2, x_3),$$

then, since

$$(x_1i + x_2j + x_3k) + (y_1i + y_2j + y_3k) = (x_1 + y_1)i + (x_2 + y_2)j + (x_3 + y_3)k,$$

we obtain the notion, familiar to the present-day reader, of vector addition,

$$\underline{x} + \underline{y} = (x_1 + y_1, x_2 + y_2, x_3 + y_3).$$

What happens to multiplication?

It turned out that physicists needed three sorts of 'products' derived from quaternion multiplication[3]. If $\mathbf{x} = x_0$ and $\mathbf{y} = y_1i + y_2j + y_3k$, then

$$\mathbf{x} \otimes \mathbf{y} = x_0y_1i + x_0y_2j + x_0y_3k$$

and we obtain the operation of 'multiplication by a scalar'. Setting $\lambda = x_0$, this has the familiar form

$$\lambda\underline{y} = (\lambda y_1, \lambda y_2, \lambda y_3).$$

Next, we observe that, if

$$\mathbf{x} = x_1i + x_2j + x_3k \text{ and } \mathbf{y} = y_1i + y_2j + y_3k,$$

then

$$\mathbf{x} \otimes \mathbf{y} = -(x_1y_1 + x_2y_2 + x_3y_3) + (x_2y_3 - x_3y_2)i + (x_3y_1 - x_1y_3)j + (x_1y_2 - x_2y_1)k,$$

and this suggests two further products, the *inner product* (called the scalar product in older texts) given by

$$\underline{x} \cdot \underline{y} = x_1y_1 + x_2y_2 + x_3y_3$$

---

[3] The key point about these products, which is proved in any respectable first course in vector calculus, is that they are 'vectorial', that is to say relations involving such products remain true when we rotate coordinate axes.

and the *cross product*

$$\underline{x} \wedge \underline{y} = (x_2 y_3 - x_3 y_2, x_3 y_1 - x_1 y_3, x_1 y_2 - x_2 y_1).$$

Exercise 11.2.1 *Show that*

$$(x_0, \underline{x}) \otimes (y_0, \underline{y}) = (x_0 y_0 - \underline{x} \cdot \underline{y}, x_0 \underline{y} + y_0 \underline{x} + \underline{x} \wedge \underline{y}).$$

Notice that we have wandered quite far from multiplication in the sense of earlier chapters. Multiplication by a scalar takes a real number λ and a vector $\underline{x}$ and produces a vector $\lambda \underline{x}$. The inner product takes two vectors $\underline{x}$ and $\underline{y}$ and produces a real number $\underline{x} \cdot \underline{y}$. The cross product does, indeed, take two vectors and produce a vector, but the formula

$$\underline{y} \wedge \underline{x} = -\underline{x} \wedge \underline{y}$$

takes some getting used to.

It must be remembered that all the fine results which adorn a first course on vector calculus were obtained by nineteenth-century mathematicians using coordinates. If you have the physical instincts of a Kelvin or a Helmholtz, you have no need of vectorial props. When Kelvin writes

> I do think however, that you would lose nothing by omitting the word 'vector' throughout. It adds nothing to the clearness or simplicity of the geometry, whether of two-dimensions or three-dimensions. Quaternions came from Hamilton after his really good work had been done; and though beautifully ingenious have been an unmixed evil to those who have touched them in any way, including Clerk Maxwell.
>
> *Quotation taken from [6]*[4]

we are listening to the man who first wrote down the result known as Stokes' theorem.

Most people now side with Maxwell rather than Kelvin and, when we move from 3 dimensions to n dimensions, vectors continue to provide a natural language. We still have addition, multiplication by a scalar, and the inner product

$$(x_1, x_2, \ldots, x_n) + (y_1, y_2 \ldots, y_n) = (x_1 + y_1, x_2 + y_2, \ldots, x_n + y_n)$$

$$\lambda(x_1, x_2, \ldots, x_n) = (\lambda x_1, \lambda x_2, \ldots, \lambda x_n)$$

$$(x_1, x_2, \ldots, x_n) \cdot (y_1, y_2 \ldots, y_n) = x_1 y_1 + x_2 y_2 + \ldots + x_n y_n.$$

Nor do we have to confine ourselves to n dimensions. Consider real valued continuous functions f with $f(t)$ defined for $0 \le t \le 1$. We can define $f + g$, λf and $f \cdot g$ by

---

[4] In addition to its many other merits, Crowe's book is a delight to those who enjoy the spectacle of very clever people being rude about other very clever people.

$$(f + g)(t) = f(t) + g(t)$$
$$(\lambda f)(t) = \lambda \times f(t)$$
$$f \cdot g = \int_0^1 f(t)g(t)\,dt.$$

Looked at in this generality, the idea of a vector space appears throughout mathematics from statistics to the study of simultaneous equations and quantum mechanics.

What about the cross product? As the switch from quaternions to vectors took place, it became clear that, at much the same time as Hamilton made his initial discovery, Grassmann had put forward a very general notion of vector space in a brilliant but almost unreadable (and certainly unread[5]) form. His treatment of the cross product allowed generalisation to many dimensions as what is now called a wedge product, but that is another story.

The quaternions are now seen as a very special system. Appendix D contains the exact statement and proof of the following theorem of Frobenius.

Theorem 11.2.2 *(Informal statement.) If $\mathbb{R}^n$ can be given the structure of a skew-field, then $n = 1$, $n = 2$ or $n = 4$. The associated structures are $\mathbb{R}$, $\mathbb{C}$ and $\mathbb{H}$.*

This theorem shows that any direct attempt to find further systems analogous to $\mathbb{C}$ is doomed to failure. The complex numbers constitute the largest system of objects that most people are content to call numbers.

As discussed in Appendix C, quaternions provide a very useful tool for the study of rotations in three dimensions and so are of particular interest to roboticists and designers of computer games.

11.3 Valedictory

The orchestra is now in place. We have caught a shadowy glimpse of the instrument makers – Egyptians, Greeks, Indians, Arabs, Germans, Frenchmen . . . and their callings – clerks, merchants, philosophers, physicists, mathematicians, ... We have heard one or two of the tunes the orchestra has played in the past, but must wait to listen to the music of the future.

---

[5] Crowe quotes a letter from Grassmann's publisher telling him that 'Your book . . . has been out of print for some time. Since your work sold hardly at all, roughly 600 copies were . . . used as waste paper and the remainder, a few odd copies, have now been sold' [6].

APPENDICES

Appendix A

Products of Many Elements

When we talked about factorisation in Theorem 4.4.2 and the results that followed, we used products like

$$x_1 x_2 \ldots x_n$$

freely. Anyone who has done Exercise 2.2.9 will agree that this is reasonable. However, we have not provided formal definitions and proofs. It is the object of this appendix to show how this can be done. We start by formalising the notion of the product of many elements.

We work with system $(\mathbb{A}, \times)$ consisting of a non-empty set $\mathbb{A}$ with an operation of multiplication $\times$ obeying the commutative law $a \times b = b \times a$ and the associative law $a \times (b \times c) = (a \times b) \times c$. However, the reader will lose nothing if she considers the special case $\mathbb{A} = \mathbb{Q}$.

Definition A.1 *If $x_j \in \mathbb{A}$, then we define $\prod_{j=1}^{n} x_j$ inductively by the rules*

$$\prod_{j=1}^{1} x_j = x_1 \text{ and } \prod_{j=1}^{n+1} x_j = \left(\prod_{j=1}^{n} x_j \right) \times x_{n+1}.$$

If the reader reflects, she will see that, when we used such products in Chapter 4, we needed the following result.

Theorem A.2 *If $y_1, y_2, \ldots, y_n$ form a rearrangement of $x_1, x_2, \ldots, x_n$, then*

$$\prod_{j=1}^{n} x_j = \prod_{j=1}^{n} y_j.$$

Exercise A.3 *Check that the result of Exercise 2.2.9 is a special case of Theorem A.2.*

233

Proof of Theorem A.2. Let $P(n)$ be the statement made by the theorem. We prove $P(n)$ for all n by induction. Observe first that $P(1)$ is the obvious statement $x_1 = x_1$ and $P(2)$ is true by the commutative law of multiplication.

Now suppose that $n \geq 2$ and $P(n)$ is true. Suppose that $b_1, b_2, \ldots, b_{n+1}$ form a rearrangement of $a_1, a_2, \ldots a_{n+1}$. We have two cases to consider.

If $a_{n+1} = b_{n+1}$, then $b_1, b_2, \ldots, b_n$ form a rearrangement of $a_1, a_2, \ldots, a_n$. By our inductive hypothesis,

$$\prod_{j=1}^{n} a_j = \prod_{j=1}^{n} b_j$$

and so, by definition,

$$\prod_{j=1}^{n+1} a_j = \left(\prod_{j=1}^{n} a_j\right) \times a_{n+1} = \left(\prod_{j=1}^{n} b_j\right) \times b_{n+1} = \prod_{j=1}^{n+1} b_j.$$

If $a_{n+1} \neq b_{n+1}$ we need a more complicated argument. Observe first that we can find $c_1, c_2, \ldots, c_{n-1}$ so that $c_1, c_2, \ldots, c_{n-1}, b_{n+1}, a_{n+1}$ form a rearrangement of $a_1, a_2, \ldots, a_{n+1}$. Now observe that $c_1, c_2, \ldots, c_{n-1}, a_{n+1}$ must form a rearrangement of $b_1, b_2, \ldots, b_n$ and so, since $P(n)$ is true,

$$\prod_{j=1}^{n} b_j = \left(\prod_{j=1}^{n-1} c_j\right) \times a_{n+1}.$$

Similarly

$$\prod_{j=1}^{n} a_j = \left(\prod_{j=1}^{n-1} c_j\right) \times b_{n+1}.$$

Thus, using the associative and commutative properties of multiplication,

$$\prod_{j=1}^{n+1} b_j = \left(\prod_{j=1}^{n} b_j\right) \times b_{n+1} = \left(\left(\prod_{j=1}^{n-1} c_j\right) \times a_{n+1}\right) \times b_{n+1}$$

$$= \left(\prod_{j=1}^{n-1} c_j\right) \times (a_{n+1} \times b_{n+1}) = \left(\prod_{j=1}^{n-1} c_j\right) \times (b_{n+1} \times a_{n+1})$$

$$= \left(\left(\prod_{j=1}^{n-1} c_j\right) \times b_{n+1}\right) \times a_{n+1} = \left(\prod_{j=1}^{n} a_j\right) \times a_{n+1} = \prod_{j=1}^{n+1} a_j.$$

Putting the two cases together, we see that we have proved that $P(n+1)$ is true and this completes the induction. ∎

We also need to know that the 'product of two products forms a product in a natural way'. (Notice that we use this result in the proof of Theorem 4.4.2.)

Lemma A.4 *If $a_j \in \mathbb{A}$, we have*

$$\left(\prod_{j=1}^{m} a_j \right) \times \left(\prod_{k=1}^{n} a_{m+k} \right) = \prod_{j=1}^{m+n} a_j.$$

Proof. This is a routine induction. Let m be fixed and let $P(n)$ be the statement that

$$\left(\prod_{j=1}^{m} a_j \right) \times \left(\prod_{k=1}^{n} a_{m+k} \right) = \prod_{j=1}^{m+n} a_j.$$

We observe that $P(1)$ follows directly from the definition

$$\left(\prod_{j=1}^{m} a_j \right) \times a_{m+1} = \prod_{j=1}^{m+1} a_j.$$

Suppose now that $P(n)$ is true. Then, using the definition of a product and the associative law for multiplication, we have

$$\left(\prod_{j=1}^{m} a_j \right) \times \left(\prod_{k=1}^{n+1} a_{m+k} \right) = \left(\prod_{j=1}^{m} a_j \right) \times \left(\left(\prod_{k=1}^{n} a_{m+k} \right) \times a_{m+n+1} \right)$$

$$= \left(\left(\prod_{j=1}^{m} a_j \right) \times \left(\prod_{k=1}^{n} a_{m+k} \right) \right) \times a_{m+n+1}$$

$$= \left(\prod_{j=1}^{m+n} a_j \right) \times a_{m+n+1} = \prod_{j=1}^{m+n+1} a_j.$$

Thus $P(n+1)$ is true and the induction is complete. ∎

Our most spectacular use of Theorem A.2 and Lemma A.4 is perhaps in the proofs of Lemmas 5.2.10 (Wilson's theorem) and Lemma 5.2.12 (ii), but Chapter 5 and our work on factorisation in Chapter 4.1 made repeated use of these results.

When we think of an operation $+$ rather than $\times$ we replace $\prod$ by $\sum$.

Exercise A.5 *Let $\mathbb{F}$ be a field. Write down the appropriate versions of Definition A.1, Theorem A.2 and Lemma A.4 for*

$$x_1 + x_2 + \ldots + x_n = \sum_{j=1}^{n} x_j,$$

with $x_j \in \mathbb{F}$.

We use the $\sum$ notation when discussing geometric series in Exercise 8.2.2, decimal expansions in Exercises 8.2.3 and 8.2.4 and elsewhere.

Here is a simple example.

Exercise A.6 *In this exercise we show how to use the ideas of the appendix to obtain the binary notation for strictly positive integers. The formula $\zeta_j \in \{0, 1\}$ just means that ζ_j takes the value 0 or 1.*

(i) If $\zeta_j \in \{0, 1\}$, show, by induction, that

$$\sum_{j=0}^{n} \zeta_j 2^j \leq 2^{n+1} - 1.$$

(ii) If $n \geq m \geq 0$, $\zeta_j \in \{0, 1\}$ for $0 \leq j \leq n$, $\eta_k \in \{0, 1\}$ for $0 \leq k \leq m$, $\zeta_n = 1$ and

$$\sum_{j=0}^{n} \zeta_j 2^j = \sum_{k=0}^{m} \eta_k 2^k,$$

show that $m = n$ and $\eta_n = 1$.

(iii) If ζ_j, $\eta_j \in \{0, 1\}$ for $0 \leq j \leq n$, and

$$\sum_{j=0}^{n} \zeta_j 2^j = \sum_{j=0}^{n} \eta_j 2^j,$$

show that $\zeta_j = \eta_j$ for $0 \leq j \leq n$.

(iv) Show that, if $2^{n+1} - 1 \geq r \geq 2^n$, then we can find $\zeta_j \in \{0, 1\}$ with $\zeta_n = 1$ such that

$$r = \sum_{j=0}^{n} \zeta_j 2^j.$$

(v) Deduce that every integer $u \geq 1$ can be written as

$$u = \sum_{j=0}^{m} \zeta_j 2^j,$$

with $\zeta_j \in \{0, 1\}$ for $0 \leq j \leq m - 1$ and $\zeta_m = 1$, in exactly one way.

(vi) *(This links with Exercise 5.5.9.) Show, by induction, that, if a is a strictly positive integer and*

$$r = \sum_{j=0}^{n} \zeta_j 2^j$$

with $\zeta_j \in \{0, 1\}$ for $0 \le j \le n$, then

$$a^r = \prod_{\zeta_j = 1} a^{2^j}.$$

(vii) *Do you think that this exercise adds anything to what is already contained in Exercise 2.1.3[1]?*

---

[1] Recall the story of the lawyer F. E. Smith, who, when a judge complained that 'I have listened to you for an hour and I am none the wiser', replied 'None the wiser, perhaps, my lord, but certainly better informed'.

Appendix B

nth Complex Roots

The object of this appendix is to provide a proof of the following result (Lemma 10.2.2) using only the methods of this book.

Lemma B.1 *If $a \in \mathbb{C}$, $a \neq 0$ and m is an integer with $m \geq 1$, then the equation*

$$az^m = |a|$$

has a solution.

Our proof runs along similar lines to that of Argand's lemma (Theorem 10.2.4) and I would strongly advise mastering the proof of that theorem before studying this appendix.

We start with some inequalities. (Note that we looked at the square root function in Lemma 7.4.12.)

Lemma B.2 *Throughout this lemma, η will be real and $|\eta| < 1/2$.*

(i) We have

$$1 - \eta^2 \leq \sqrt{1 - \eta^2} \leq 1.$$

(ii) For each integer $m \geq 1$, there exists a $C_m > 0$ depending on m such that

$$\left| \left(\sqrt{1 - \eta^2} + i\eta \right)^m - 1 - im\eta \right| \leq C_m \eta^2$$

whenever $|\eta| \leq 1/4$.

Proof. [If the reader has some experience with this sort of thing she will be able to produce a much quicker proof of (ii).]

(i) Since $0 < 1 - \eta^2 \leq 1$,

$$(1 - \eta^2)^2 \leq (1 - \eta^2) \leq 1,$$

239

so (see Exercise 7.4.13)

$$1 - \eta^2 \leq \sqrt{1 - \eta^2} \leq 1.$$

(ii) Set $u = (\sqrt{1 - \eta^2} + i\eta) - 1$. By Exercise 10.1.14, we can find an $A_m > 0$ such that

$$|(1 + u)^m - 1 - mu| \leq A_m |u|^2$$

whenever $|u| \leq 1/2$.

Now

$$|u|^2 = (\sqrt{1 - \eta^2} - 1)^2 + \eta^2 = (1 - \eta^2) - 2\sqrt{1 - \eta^2} + 1 + \eta^2$$
$$= 2(1 - \sqrt{1 - \eta^2}) \leq 4\eta^2$$

so $|u| \leq 2|\eta|$ and we have

$$|(1 + u)^m - 1 - mu| \leq A_m |u|^2 \leq 4 A_m \eta^2$$

whenever $|\eta| \leq 1/4$.

We also have

$$\left|(1 - mu) - (1 - im\eta)\right| = m|u - i\eta| = m\left(1 - \sqrt{1 - \eta^2}\right)$$
$$\leq m\left(1 - (1 - \eta^2)\right) = m\eta^2$$

so, combining the results of this paragraph and the previous one,

$$\left|\left(\sqrt{1 - \eta^2} + i\eta\right)^m - 1 - im\eta\right| \leq |(1 + u)^m - 1 - mu|$$
$$+ \left|(1 - mu) - (1 - im\eta)\right|$$
$$\leq 4 A_m \eta^2 + m\eta^2 = (4 A_m + m)\eta^2$$

whenever $|\eta| \leq 1/4$. Setting $C_m = 4 A_m + m$, we have the desired result. ∎

We can now prove Lemma B.1.

Informal proof of Lemma B.1. Our formal proof of the lemma is obscured by notation. Informally, we observe that, if we write $g(z) = \Re(az^m)$ then g has a maximum at some point $w = x + iy$ on the circle Γ defined by $|z| = 1$. If η is real,

$$z_\eta = w\left(\sqrt{1 - \eta^2} + i\eta\right)$$

also lies on Γ and, if η is small, lies close to w. In fact, if η is small,

$$z_\eta^m = w^m (\sqrt{1 - \eta^2} + i\eta)^m$$
$$= w^m \big(1 + i\eta + O(\eta^2)\big)^m = w^m \big(1 + mi\eta + O(\eta^2)\big),$$

so

$$\Re(az^m) = \Re(aw^m) - \eta\Im(aw^m) + O(\eta^2),$$

where $O(\eta^2)$ is to be read as 'with an error no more than $A\eta^2$ for some A'. Unless aw^m is real we can now choose some small real (positive or negative) η such that $\Re(az_\eta^m) > \Re(aw^m)$.

Thus aw^m is real so (since $|w| = 1$) $aw^m = \pm|a|$. If $aw^m = -|a|$ then, since $g(w)$ is a maximum, $g(z) = -|a|$ for all z with $|z| = 1$. Since this is impossible, $aw^m = |a|$. $\blacksquare$

Formal proof of Lemma B.1. We know that the function $f : \mathbb{C} \to \mathbb{C}$ given by $f(z) = az^m$ is a polynomial and so is continuous. If we take $g(z) = \Re f(z)$, that is to say we take $g(z)$ to be the real part of $f(z)$ (see page 190), we know that

$$|g(z) - g(w)| = \big|\Re\big(f(z) - f(w)\big)\big| \le |f(z) - f(w)|$$

so $g : \mathbb{C} \to \mathbb{R}$ is continuous.

By Exercise 9.3.8, we can find a $w \in \mathbb{C}$ such that $|w| = |a|$ and

$$g(w) \ge g(z)$$

for all z with $|z| = |a|$. We write $g(w) = x + iy$ with x and y real.

By Lemma B.2, there exists a $C > 0$ such that

$$C\eta^2 \ge \left|\left(\sqrt{1 - \eta^2} + i\eta\right)^m - 1 - im\eta\right|$$

whenever η is real and $|\eta| \le 1/4$. Since $|\Re v| \le |v|$, we have, in particular,

$$C\eta^2 \ge \left|\Re\left(\left(\sqrt{1 - \eta^2} + i\eta\right)^m - 1 - im\eta\right)\right|$$
$$= \left|\Re\left((\sqrt{1 - \eta^2} + i\eta)^m\right) - 1\right|$$

and, since $|\Im v| \le |v|$,

$$C\eta^2 \ge \left|\Im\left(\left(\sqrt{1 - \eta^2} + i\eta\right)^m - 1 - im\eta\right)\right|$$
$$= \left|\Im\left((\sqrt{1 - \eta^2} + i\eta)^m\right) - m\eta\right|$$

for all η real and $|\eta| \le 1/4$.

From now on, we suppose that the conditions on η just stated are satisfied and set

$$z_\eta = w(\sqrt{1-\eta^2}+i\eta).$$

We observe that $|z_\eta| = |w|$ and (by a simple induction, if necessary)

$$f(z_\eta) = a\big(w(\sqrt{1-\eta^2}+i\eta)\big)^m = f(w)(\sqrt{1-\eta^2}+i\eta)^m$$

so that

$$
\begin{aligned}
g(z_\eta) &= \Re\big(f(w)(\sqrt{1-\eta^2}+i\eta)^m\big) \\
&= \Re\big((x+iy)(\sqrt{1-\eta^2}+i\eta)^m\big) \\
&= x\Re\big((\sqrt{1-\eta^2}+i\eta)^m\big) - y\Im\big((\sqrt{(1-\eta^2)}+i\eta)^m\big).
\end{aligned}
$$

Thus, using the results of the previous paragraph,

$$|g(z_\eta) - x + my\eta| \le (|x|+|y|)C\eta^2 \le 2|a|C\eta^2.$$

Since $g(w) = x$, this gives

$$|g(z_\eta) - g(w) + my\eta| \le (|x|+|y|)C\eta^2 \le 2|a|C\eta^2.$$

Suppose that $y > 0$. Observing that all the terms in the final inequality are real, we now have

$$g(z_\eta) \ge g(w) - my\eta - 2|a|C\eta^2.$$

Choosing η with $0 > \eta > -(2|a|C)/my$ gives $g(z_\eta) > g(w)$, contrary to our definition of w. If $y < 0$, a similar argument (left to the reader) gives the same contradiction. Thus we must have $y = 0$ and $g(w)$ real.

We have shown that $g(w) = |a|$ or $g(w) = -|a|$. If $g(w) = -|a|$, then, since g takes its maximum value at w, and $g(z) \ge -|a|$ for all z with $|z| = 1$, it follows that $g(z) = -|a|$ and so $f(z) = -|a|$ for all z with $|z| = 1$. Since a polynomial of degree m cannot take the same value at more than m points, it follows that $g(w) \ne -|a|$. Thus $g(w) = |a|$, and so $aw^m = |a|$. ∎

Exercise B.3 *Prove the result left to the reader in the second to last paragraph of the preceding proof.*

Appendix C
How Do Quaternions Represent Rotations?

The contents of this section are easy, but we shall require results from courses which use vectors to study $\mathbb{R}^3$. The reader will need to know the geometric meaning of the inner product $\underline{a} \cdot \underline{b}$ and the cross product $\underline{a} \wedge \underline{b}$ where $\underline{c}$ denotes a vector in $\mathbb{R}^3$. She will also need to know the result, going back to Euler, which states that every rotation in $\mathbb{R}^3$ has an axis. (When we speak of a rotation, we will assume that it leaves the origin $\underline{0}$ fixed.)

We write quaternions in the form suggested in Exercise 11.1.2 (i),

$$\mathbf{x} = (x_0, \underline{x}) = (x_0, x_1, x_2, x_3)$$

so $\underline{x}$ is a vector in $\mathbb{R}^3$. We write $\|\underline{x}\| = \sqrt{x_1^2 + x_2^2 + x_3^2}$. If $\|\underline{x}\| = 1$, we say that $\underline{x}$ has unit length. We wish to prove the following result.

Theorem C.1 *If $\underline{q}$ has unit length, θ is a real number and we form the quaternion*

$$\mathbf{q} = \left(\cos \frac{\theta}{2}, \sin \frac{\theta}{2} \underline{q} \right),$$

then, whenever $\underline{x} \in \mathbb{R}^3$,

$$\mathbf{q} \otimes (0, \underline{x}) \otimes \mathbf{q}^{-1} = (0, \underline{y})$$

where $\underline{y}$ is the rotation of $\underline{x}$ about the axis $\underline{q}$ by an angle θ.

Our proof of Theorem C.1 is a long calculation using the formula

$$(a_0, \underline{a}) \otimes (b_0, \underline{b}) = (a_0 b_0 - \underline{a} \cdot \underline{b}, a_0 \underline{b} + b_0 \underline{a} + \underline{a} \wedge \underline{b}),$$

together with lots of elementary results on vector geometry for $\mathbb{R}^3$.

Exercise C.2 *Suppose that $\underline{q}$ and $\underline{n}$ are perpendicular unit vectors and $\theta \in \mathbb{R}$. Recall or prove the following results.*

(i) $q \cdot q = 1$, $q \cdot n = 0$, $q \wedge n = -n \wedge q$ and $q \wedge q = 0$.

(ii) q, n and $q \wedge n$ *are mutually orthogonal unit vectors. In particular, we have* $(q \wedge n) \cdot q = 0$.

(iii) $(q \wedge n) \wedge q = n$.

(iv) $2(\cos\theta/2)(\sin\theta/2) = \sin\theta$, $(\cos\theta/2)^2 - (\sin\theta/2)^2 = \cos\theta$.

Proof of Theorem C.1. By using Exercise 11.1.7 or direct computation, we check that

$$\mathbf{q}^{-1} = \left(\cos\frac{\theta}{2}, -\sin\frac{\theta}{2}q \right).$$

We first consider the case when $x = q$. We then have

$$
\begin{aligned}
(\mathbf{q} \otimes (0, q)) \otimes \mathbf{q}^{-1} &= \big(((\cos\theta/2), (\sin\theta/2)q) \otimes (0, q)\big) \otimes \mathbf{q}^{-1} \\
&= (-(\sin\theta/2), (\cos\theta/2)q) \otimes ((\cos\theta/2), -(\sin\theta/2)q) \\
&= (0, ((\sin\theta/2)^2 + (\cos\theta/2)^2)q) = (0, q).
\end{aligned}
$$

Next we look at the case when $x = n$ a unit vector perpendicular to q. We then have

$$
\begin{aligned}
(\mathbf{q} \otimes (0, n)) \otimes \mathbf{q}^{-1} &= \big(((\cos\theta/2), (\sin\theta/2)q) \otimes (0, n)\big) \otimes \mathbf{q}^{-1} \\
&= (0, (\cos\theta/2)n + (\sin\theta/2)q \wedge n) \otimes ((\cos\theta/2), -(\sin\theta/2)q) \\
&= (0, ((\cos\theta/2)^2 - (\sin\theta/2)^2)n + 2(\cos\theta/2)(\sin\theta/2)q \wedge n) \\
&= (0, (\cos\theta)n + (\sin\theta)q \wedge n).
\end{aligned}
$$

Now consider a general vector x. We can find a unit vector n perpendicular to q together with real numbers α and β such that

$$x = \alpha q + \beta n.$$

Using the previous paragraph,

$$
\begin{aligned}
\mathbf{q} \otimes (0, x) \otimes \mathbf{q}^{-1} &= \mathbf{q} \otimes (0, \alpha q + \beta n) \otimes \mathbf{q}^{-1} \\
&= \alpha(\mathbf{q} \otimes (0, q)) \otimes \mathbf{q}^{-1} + \beta(\mathbf{q} \otimes (0, n)) \otimes \mathbf{q}^{-1} \\
&= (0, \alpha q + \beta(\cos\theta)n + \beta(\sin\theta)q \wedge n) = (0, y),
\end{aligned}
$$

where y is the vector obtained by rotating x through an angle θ about the axis q. ∎

Exercise C.3 *Suppose that p is a non-zero quaternion. Show that the formula*

$$\mathbf{p} \otimes (0, x) \otimes \mathbf{p}^{-1} = (0, y)$$

defines a rotation taking $\underline{x}$ to $\underline{y}$. What is the axis of the rotation and by what angle does it rotate?

Remark The contents of this appendix show that the kind of over-bracketing adopted in this book can be positively harmful. An expression like $\mathbf{p} \otimes (0, \underline{x}) \otimes \mathbf{p}^{-1}$ should be considered as a single item and not artificially cut up as $\left(\mathbf{p} \otimes (0, \underline{x})\right) \otimes \mathbf{p}^{-1}$ or $\mathbf{p} \otimes \left((0, \underline{x}) \otimes \mathbf{p}^{-1}\right)$. We say that $(0, \underline{x})$ is *conjugated* by $\mathbf{p}$.

Exercise C.4 *Suppose that $\mathbf{q}_j$ is a unit quaternion corresponding to a rotation R_j in the manner described above. Show that $\mathbf{q}_2 \otimes \mathbf{q}_1$ corresponds to the rotation R_1 followed by the rotation R_2.*

The quaternionic representation of rotation is a particularly natural one because, given the appropriate quaternion, we can read off the rotation axis and rotation angle. It comes into its own when we wish to find the effect of many successive rotations, since then, as Exercise C.3 shows, we only need to multiply the appropriate quaternions.

Exercise C.5 *(Only if you know about matrices and only if you are interested in computation.)*

(i) *Check that the standard method for multiplying 3×3 matrices requires 27 multiplications and 18 additions (because we use negative numbers, addition and subtraction are not distinguished). Check that the obvious method for multiplying two quaternions requires 16 multiplications and 12 additions.*

(ii) *Observe that, if a unit quaternion $\mathbf{p}$ is slightly perturbed to $\mathbf{q}$, then $\mathbf{q}/\|\mathbf{q}\|$ is a unit quaternion which is close to $\mathbf{p}$. Suppose that Q a rotation matrix is slightly perturbed to obtain a matrix P. Think about the problem of using P to find a rotation matrix R close to Q.*

[This exercise merely indicates why quaternionic methods might be useful, it does not state that they will be useful.]

Exercise C.6 *Let $\mathbf{p}$ be a non-zero quaternion. Show that the function $f : \mathbb{H} \to \mathbb{H}$ given by $f(\mathbf{u}) = \mathbf{p} \otimes \mathbf{u} \otimes \mathbf{p}^{-1}$ is a skew-field isomorphism with $f(\mathbf{u}^*) = f(\mathbf{u})^*$.*
Show, in particular, that if we set $\hat{i} = f(i)$, $\hat{j} = f(j)$ and $\hat{k} = f(k)$, then

$$\hat{i}^2 = \hat{j}^2 = \hat{k}^2 = \hat{i} \otimes \hat{j} \otimes \hat{k} = -1.$$

Appendix D

Why Are the Quaternions So Special?

The object of this final appendix is to give an exact statement and proof of Theorem 11.2.2. The result is due to Frobenius, and I shall follow the proof given by Palais [27].

I have put this material in an appendix because it requires the knowledge of vector spaces contained in a first university course in abstract algebra. However, the proof of our first theorem only requires the reader to understand the statement of the following lemma and accept its truth.

Lemma D.1 *If U is a vector space of dimension n over $\mathbb{R}$, then, if $\mathbf{u}_0, \mathbf{u}_1, \ldots, \mathbf{u}_n \in U$, we can find $\lambda_0, \lambda_1, \ldots, \lambda_n \in \mathbb{R}$, not all zero, such that*

$$\lambda_0 \mathbf{u}_0 + \lambda_1 \mathbf{u}_1 + \cdots + \lambda_n \mathbf{u}_n = \mathbf{0}.$$

This result just stated becomes particularly plausible if we think in terms of $U = \mathbb{R}^n$.

Theorem D.2 *Consider the system $(U, +, \otimes, \mathbb{R})$. Suppose that $(U, +, \mathbb{R})$ is a finite-dimensional vector space over $\mathbb{R}$, that $(U, +, \otimes)$ is a field and that the multiplicative unit $\underline{e}$ of $(U, +, \otimes)$ satisfies the condition*

$$(\lambda \underline{e}) \otimes \underline{u} = \lambda \underline{u}$$

for all $\lambda \in \mathbb{R}$ and $\underline{u} \in U$.

Either the vector space U is one-dimensional and the field U is field isomorphic to $\mathbb{R}$ or the vector space U is two-dimensional and the field U is field isomorphic to $\mathbb{C}$.

Proof. Observe that $f : \mathbb{R} \to U$ given by $f(\lambda) = \lambda \underline{e}$ is an injective function which preserves the field operations. We identify the one-dimensional subspace generated by $\underline{e}$ with $\mathbb{R}$ and write $\lambda = \lambda \underline{e}$. In particular, $\underline{e} = 1$ and $\underline{0} = 0$.

Notice, for future use that, since U is a field, the equation $\underline{u}^2 + \underline{e} = 0$ either has no solution or has exactly two solutions which we denote by $\underline{i}$ and $-\underline{i}$.

If U is one-dimensional, we are done. If not, then, thinking of U as vector space, there are elements of U which are not scalar multiples of $\underline{e}$. Let $\underline{d}$ be such an element. If U has dimension n, then applying Lemma D.1 to the vectors $\underline{e}$, $\underline{d}$, $\underline{d}^2$, ..., $\underline{d}^n$ (obtained by taking powers in the field U), we know that there exist $\lambda_0, \lambda_1, \ldots, \lambda_n \in \mathbb{R}$ not all zero such that

$$\lambda_0 \underline{e} + \lambda_1 \underline{d} + \cdots + \lambda_n \underline{d}^n = \underline{0}.$$

Thus there exists an m with $0 \le m \le n$, $\lambda_m \ne 0$ and

$$\lambda_0 \underline{e} + \lambda_1 \underline{d} + \cdots + \lambda_m \underline{d}^m = \underline{0}.$$

Since $\underline{d}$ is not a scalar multiple of $\underline{e}$ we have $m \ge 1$.

Set

$$P(t) = \lambda_0 + \lambda_1 t + \cdots + \lambda_m t^m.$$

By the fundamental theorem of algebra [see, in particular Exercise 10.2.7 (vi)], working over $\mathbb{R}$,

$$P(t) = Q_1(t) Q_2(t) \ldots Q_r(t)$$

where the Q_j are linear or quadratic polynomials with real coefficients. Thus

$$P(t\underline{e}) = Q_1(t\underline{e}) \otimes Q_2(t\underline{e}) \otimes \cdots \otimes Q_r(t\underline{e})$$

for all $t \in \mathbb{R}$. By Exercise 10.1.8, this means that

$$P(\underline{u}) = Q_1(\underline{u}) \otimes Q_2(\underline{u}) \otimes \cdots \otimes Q_r(\underline{u})$$

for all $\underline{u} \in U$. Thus

$$Q_1(\underline{d}) \otimes Q_2(\underline{d}) \otimes \cdots \otimes Q_r(\underline{d}) = P(\underline{d}) = \underline{0}.$$

Since U is a field, we must have $Q_j(\underline{d}) = \underline{0}$ for some j. We cannot have Q_j linear, since this would make $\underline{d}$ a scalar multiple of $\underline{e}$. Thus Q_j is quadratic and we can find a, b, $c \in \mathbb{R}$ such that

$$a\underline{d}^2 + b\underline{d} + c\underline{e} = \underline{0}$$

with $a \ne 0$. By multiplying through by a^{-1}, we may suppose that $a = 1$ and

$$\underline{d}^2 + b\underline{d} + c\underline{e} = \underline{0}.$$

Recalling that we identify e with 1, completing the square yields

$$(d + b/2)^2 = (b^2 - 4c)/4.$$

If $(b^2 - 4c) \geq 0$, then $d = (-b \pm \sqrt{b^2 - 4c})/2 \in \mathbb{R}$ which we have excluded. Thus $b^2 - 4c < 0$ and we have

$$\left(\frac{2d + b}{\sqrt{4c - b^2}} \right)^2 + 1 = 0.$$

We have shown that U is a field in which $u^2 + 1 = 0$ (that is to say, $u^2 + e = 0$) has a root. Recalling that we have agreed to call the two roots i and $-i$, we see that

$$d = Ae + Bi$$

for appropriate $A, B \in \mathbb{R}$.

We have shown that every element u of U may be written as

$$u = xe + yi$$

with $x, y \in \mathbb{R}$. It follows that the function $g : \mathbb{C} \to U$ defined by $g(x + iy) = xe + yi$ is a bijection. A simple check shows that g is a field isomorphism. ∎

Exercise D.3 *Do the simple check just mentioned.*

Theorem D.2 shows why Hamilton had to abandon one of the field laws. Our final theorem explains why, even if we jettison the commutative law of multiplication, the only additional object we obtain is the quaternion algebra. The proof requires substantially more abstract algebra.

Theorem D.4 *Consider the system* $(V, +, \otimes, \mathbb{R})$. *Suppose that* $(V, +, \mathbb{R})$ *is a finite-dimensional vector space over* $\mathbb{R}$, *that* $(V, +, \otimes)$ *is a skew-field and that the multiplicative unit* e *of* $(V, +, \otimes)$ *satisfies the condition*

$$(\lambda e) \otimes u = \lambda u = u \otimes (\lambda e)$$

for all $\lambda \in \mathbb{R}$ *and* $u \in V$.

Either the vector space V *is one-dimensional and* V *is a field isomorphic to* $\mathbb{R}$ *or the vector space* V *is two-dimensional and* V *is a field isomorphic to* $\mathbb{C}$ *or* V *is four-dimensional and is skew-field isomorphic to* $\mathbb{H}$

Notice how much more detail we need to put in compared with our informal statement in Theorem 11.2.2. We are dealing not with sets but with *structures* and how those structures interact.

Exercise D.5 *We use the notation and hypotheses of Theorem D.4. Show that*

$$\lambda(\underline{v} \otimes \underline{w}) = (\lambda \underline{v}) \otimes \underline{w} = \underline{v} \otimes (\lambda \underline{w})$$

for all $\lambda \in \mathbb{R}$ *and* $\underline{v}, \underline{w} \in V$.

The key to the proof of Theorem D.4 is the following lemma.

Lemma D.6 *Consider the system* $(V, +, \otimes, \mathbb{R})$. *Suppose that* $(V, +, \mathbb{R})$ *is a finite-dimensional vector space over* $\mathbb{R}$, *that* $(V, +, \otimes)$ *is a skew-field and that the multiplicative unit* $\underline{e}$ *of* $(V, +, \otimes)$ *satisfies the condition*

$$(\lambda \underline{e}) \otimes \underline{v} = \lambda \underline{v} = \underline{v} \otimes (\lambda \underline{e})$$

for all $\lambda \in \mathbb{R}$ *and* $\underline{v} \in V$.

Suppose that $\underline{d} \in V$ *and there is no* $x \in \mathbb{R}$ *such that* $\underline{d} = x \underline{e}$, *Then*

$$E = \{x\underline{e} + y\underline{d} : x, y \in \mathbb{R}\}$$

is a sub-skew-field with commutative multiplication (and so, by Theorem D.2, a field isomorphic to $\mathbb{C}$). *Further* E *consists precisely of those* $\underline{u} \in V$ *with* $\underline{u} \otimes \underline{d} = \underline{d} \otimes \underline{u}$.

Proof. We consider the subspace E of all elements $x\underline{e} + y\underline{d}$ with $x, y \in \mathbb{R}$ and observe that, if $\underline{z} = x\underline{e} + y\underline{d}$ and $\underline{z}' = x'\underline{e} + y'\underline{d}$, then

$$\underline{z} \otimes \underline{z}' = xx'\underline{e} + (xy' + yx')\underline{d} + yy'\underline{d}^2 = x'x\underline{e} + (x'y + y'x)\underline{d} + y'y\underline{d}^2 = \underline{z}' \otimes \underline{z}.$$

In other words, every element of E commutes multiplicatively with every other.

Since V is finite-dimensional, there must be a subspace F of largest dimension containing E such that every element of F commutes multiplicatively with every other[1]. We claim that any $\underline{v} \in V$ which commutes multiplicatively with every element of F actually lies in F. To see this, observe that, if F' is the vector subspace of elements $\underline{f} + y\underline{v}$ with $\underline{f} \in F$ and $y \in \mathbb{R}$, then essentially the same computation as in the previous paragraph shows that any element of F' commutes multiplicatively with every other. Since F has maximal dimension and F' contains F, we must have $F' = F$ and $\underline{v} \in F$.

We use the fact that any $\underline{v} \in V$ which commutes multiplicatively with every element of F actually lies in F to show that F is a sub-skew-field of U. Observe that, if $\underline{a}, \underline{b} \in F$, then, using the two-sided distributive law,

$$\underline{f} \otimes (\underline{a} + \underline{b}) = (\underline{f} \otimes \underline{a}) + (\underline{f} \otimes \underline{b}) = (\underline{a} \otimes \underline{f}) + (\underline{b} \otimes \underline{f}) = (\underline{a} + \underline{b}) \otimes \underline{f}$$

---

[1] Note that, at this point, we do not know that there is only one such subspace.

for all $\underset{\sim}{f} \in F$ and so $\underset{\sim}{a} + \underset{\sim}{b} \in F$. The same kind of argument shows that, if $\underset{\sim}{a}, \underset{\sim}{b} \in F$, then $\underset{\sim}{a} \otimes \underset{\sim}{b} \in F$.

We must also show that multiplicative inverses behave correctly. Suppose that $\underset{\sim}{a} \in F$ and $\underset{\sim}{a} \neq 0$. A standard algebraic manipulation shows that

$$\underset{\sim}{a}^{-1} \otimes \underset{\sim}{f} = (\underset{\sim}{a}^{-1} \otimes \underset{\sim}{f}) \otimes \underset{\sim}{e} = (\underset{\sim}{a}^{-1} \otimes \underset{\sim}{f}) \otimes (\underset{\sim}{a} \otimes \underset{\sim}{a}^{-1})$$

$$= (\underset{\sim}{a}^{-1} \otimes (\underset{\sim}{f} \otimes \underset{\sim}{a})) \otimes \underset{\sim}{a}^{-1} = (\underset{\sim}{a}^{-1} \otimes (\underset{\sim}{a} \otimes \underset{\sim}{f})) \otimes \underset{\sim}{a}^{-1}$$

$$= ((\underset{\sim}{a}^{-1} \otimes \underset{\sim}{a}) \otimes \underset{\sim}{f})) \otimes \underset{\sim}{a}^{-1} = (\underset{\sim}{e} \otimes \underset{\sim}{f}) \otimes \underset{\sim}{a}^{-1} = \underset{\sim}{f} \otimes \underset{\sim}{a}^{-1}$$

for all $\underset{\sim}{f} \in F$ and so $\underset{\sim}{a}^{-1} \in F$.

If we consider F as a stand-alone object, we see that F is a vector space of dimension greater than 1. Since F is a field and a vector space of dimension greater than 1, Theorem D.2 tells us that F has dimension 2. Since $F \supseteq E$ and E has dimension 2 we have $E = F$ and we are done. ∎

Exercise D.7

(i) *Perform the 'essentially the same' calculation referred to in the third paragraph of our proof to show that every element of F' commutes multiplicatively with every other.*

(ii) *Prove the statement made at the end of the fourth paragraph of the proof above that, if $\underset{\sim}{a}, \underset{\sim}{b} \in F$, then $\underset{\sim}{a} \otimes \underset{\sim}{b} \in F$.*

The next two exercises are easy but intended to give some insight into the proof that follows.

Exercise D.8 *Consider the quaternions $\mathbb{H}$. We write E^+ for the set of quaternions $x + y\mathbf{i}$ with x and y real and E^- for the set of quaternions $E^+ \otimes \mathbf{j}$, that is to say, we take E^- to be the collection of quaternions $\mathbf{u} \otimes \mathbf{j}$ with $\mathbf{u} \in E^+$.*

(i) *Identify E^-. Check that E^+ and E^- are subspaces of dimension 2.*

(ii) *Show that $\mathbf{a} \otimes \mathbf{i} = \mathbf{i} \otimes \mathbf{a}$ if and only if $\mathbf{a} \in E^+$ and that $\mathbf{b} \otimes \mathbf{i} = -\mathbf{i} \otimes \mathbf{b}$ if and only if $\mathbf{b} \in E^-$.*

(iii) *Show that $\mathbb{H}$ is the direct sum of E^+ and E^-, that is to say, $E^+ \cap E^- = \{0\}$ and every $\mathbf{u} \in \mathbb{H}$ can be written as $\mathbf{u} = \mathbf{a} + \mathbf{b}$ with $\mathbf{a} \in E^+$ and $\mathbf{b} \in E^-$.*

(iv) *Show that, if $\mathbf{u} \in E^-$, then $\mathbf{u} \otimes \mathbf{u}$ is real and, if $\mathbf{u} \neq 0$, then $\mathbf{u} \otimes \mathbf{u} < 0$.*

Exercise D.9 *We work with the notation and conditions of Theorem D.4. Consider $\underset{\sim}{u}, \underset{\sim}{v} \in V$ such that $\underset{\sim}{u} \otimes \underset{\sim}{v} = -\underset{\sim}{v} \otimes \underset{\sim}{u}$ and $\underset{\sim}{u} \otimes \underset{\sim}{u} = \underset{\sim}{v} \otimes \underset{\sim}{v} = -1$*

(i) If $q = c\underline{u} + d\underline{v}$ with c, $d \in \mathbb{R}$, show that

$$c\underline{v} = -\frac{1}{2}(q \otimes \underline{u} + \underline{u} \otimes q).$$

Deduce that

$$c\underline{u} = \frac{1}{2}(q + \underline{u} \otimes q \otimes \underline{u}).$$

Find a similar expression for $d\underline{v}$ in terms of q and $\underline{u}$.

(ii) Show that if a, $b \in \mathbb{R}$ and $a\underline{u} + b\underline{v} = \underline{0}$, then $a = b = 0$ (that is to say, $\underline{u}$ and $\underline{v}$ are linearly independent).

We are now in a position to prove Theorem D.4, but the reader is warned, once again, that our proof involves concepts and notations which, though part of elementary abstract algebra, have not been dealt with in this book.

Proof of Theorem D.4. If V has dimension 1, then by inspection V is field isomorphic to $\mathbb{R}$. From now on we assume that V has dimension greater than 1.

By Lemma D.6, this means that V contains a sub-skew-field E which is a field isomorphic to $\mathbb{C}$. In particular it contains an element $\underline{i}$ with $\underline{i}^2 = -\underline{1}$. We write $D^+ = E$ observing that D^+ is the set of q such that $q \otimes \underline{i} = \underline{i} \otimes q$ (that is to say, the set of all elements which commute under multiplication with $\underline{i}$) and take D^- to be the set of q such that $q \otimes \underline{i} = -\underline{i} \otimes q$ (that is to say, the set of all elements which 'anti-commute' with $\underline{i}$). We note that D^+ and D^- are vector subspaces of V.

If $q \in D^+ \cap D^-$ we have

$$q \otimes \underline{i} = \underline{i} \otimes q = -q \otimes \underline{i},$$

so $q \otimes \underline{i} = \underline{0}$ and, since $\underline{i} \neq \underline{0}$ and we are in a skew-field, $q = \underline{0}$. Thus $D^+ \cap D^- = \{\underline{0}\}$.

On the other hand, direct calculation shows that, whenever $q \in V$,

$$\underline{i} \otimes (q - \underline{i} \otimes q \otimes \underline{i}) = \underline{i} \otimes q + q \otimes \underline{i} = (q - \underline{i} \otimes q \otimes \underline{i}) \otimes \underline{i}$$

and

$$\underline{i} \otimes (q + \underline{i} \otimes q \otimes \underline{i}) = \underline{i} \otimes q - q \otimes \underline{i} = -(q + \underline{i} \otimes q \otimes \underline{i}) \otimes \underline{i}.$$

Thus

$$\frac{1}{2}(q - \underline{i} \otimes q \otimes \underline{i}) \in D^+, \quad \frac{1}{2}(q + \underline{i} \otimes q \otimes \underline{i}) \in D^-.$$

Since

$$\frac{1}{2}(q - \underline{i} \otimes q \otimes \underline{i}) + \frac{1}{2}(q + \underline{i} \otimes q \otimes \underline{i}) = q,$$

we have shown that V is the direct sum of D^+ and D^-. It follows, by a standard theorem, that

$$\dim V = \dim D^+ + \dim D^-,$$

where $\dim F$ denotes the dimension of F.

We know that $\dim D^+ = 2$. If $\dim D^- = 0$, then $D^- = \{0\}$ and $V = D^+$, so V has dimension 2 and is field isomorphic to $\mathbb{C}$. From now on, we suppose that $\dim D^- \neq 0$ and so, in particular, there exists a non-zero $\underline{p} \in D^-$.

If $\underline{q} \in D^+$, we have

$$(\underline{q} \otimes \underline{p}) \otimes \underline{i} = \underline{q} \otimes (\underline{p} \otimes \underline{i}) = -\underline{q} \otimes (\underline{i} \otimes \underline{p})$$
$$= -(\underline{q} \otimes \underline{i}) \otimes \underline{p} = -(\underline{i} \otimes \underline{q}) \otimes \underline{p}$$
$$= -\underline{i} \otimes (\underline{q} \otimes \underline{p}).$$

Thus we may define a function $T : D^+ \to D^-$ by $T(\underline{q}) = \underline{q} \otimes \underline{p}$. We observe that T is a linear function.

We note that, since $\underline{p} \in D^-$, it follows (using the 'standard algebraic manipulation' in the proof of Lemma D.6) that $\underline{p}^{-1} \in D^-$ and an argument along the lines of the previous paragraph shows that $S(\underline{k}) = \underline{k} \otimes \underline{p}^{-1}$ defines a linear function $S : D^- \to D^+$. Since $ST : D^+ \to D^+$ and $TS : D^- \to D^-$ are identity functions, T is an isomorphism. Thus $\dim D^- = \dim D^+$ and $\dim V = 4$.

We have shown why it is impossible to produce a 'quaternionic system' unless we work in dimension 4. We still have to prove that there is only one quaternionic system, that is to say, in the formal language of this theorem, that V is a skew-field isomorphic to $\mathbb{H}$.

To do this, we look at $\underline{p}$ and $\underline{p}^2$. Since

$$\underline{i} \otimes \underline{p}^2 = (\underline{i} \otimes \underline{p}) \otimes \underline{p} = (-\underline{p} \otimes \underline{i}) \otimes \underline{p} = -\underline{p} \otimes (\underline{i} \otimes \underline{p}) = \underline{p} \otimes (\underline{p} \otimes \underline{i}) = \underline{p}^2 \otimes \underline{i}$$

we know that $\underline{p}^2 \in D^+$. However, Lemma D.6 tells us that the set F of elements $x\underline{e} + y\underline{p}$ with $x, y \in \mathbb{R}$ form a field and so $\underline{p}^2 \in F$ and

$$\underline{p}^2 \in D^+ \cap F.$$

Since $\underline{p} \notin D^+$, $D^+ \cap F$ is the one dimensional space of elements of the form $x\underline{e}$ with $x \in \mathbb{R}$ and, in particular, $\underline{p}^2 = c\underline{e}$ for some real c.

We cannot have $c = 0$, so $c > 0$ or $c < 0$. If $c > 0$, then $c = d^2$ for some $d \in \mathbb{R}$ with $d \neq 0$. It follows that the three distinct elements $d\underline{e}$, $-d\underline{e}$ and $\underline{p}$ of F are all solutions of

$$\underline{a}^2 - c\underline{e} = \underline{0}.$$

Since a polynomial of degree n in a field can have at most n roots (see Theorem 10.1.7), this is impossible. Thus $c < 0$ and $c = -d^2$ for some $d \in \mathbb{R}$ with $d \neq 0$. If we set $\underline{j} = d^{-1}\underline{p}$, we see that $\underline{j} \in D^-$ and $\underline{j}^2 = -1$.

The rest of the proof is plain sailing. We set $\underline{k} = \underline{i} \otimes \underline{j}$ and observe that (since $\underline{j} \in D^-$) we must have $\underline{j} \otimes \underline{i} = -\underline{k}$. Simple calculations now give

$$\underline{k}^2 = -(\underline{j} \otimes \underline{i}) \otimes (\underline{i} \otimes \underline{j}) = -(\underline{j} \otimes \underline{i}^2) \otimes \underline{j} = \underline{j}^2 = -1$$

and

$$\underline{j} \otimes \underline{k} = \underline{j} \otimes (\underline{i} \otimes \underline{j}) = -\underline{j} \otimes (\underline{j} \otimes \underline{i}) = -\underline{j}^2 \otimes \underline{i} = \underline{i}$$

Similarly, $\underline{k} \otimes \underline{j} = -\underline{i}$ and $\underline{k} \otimes \underline{i} = -\underline{i} \otimes \underline{k} = \underline{j}$.

If we consider D^+ and D^- as vector spaces over $\mathbb{R}$, then $\underline{e}$ and $\underline{i}$ form a basis for D^+ and $\underline{j}$ and $\underline{k}$ form a basis for D^- (recall that D^- has dimension 2 and note that, since $\underline{j}$ and $\underline{k}$ do not commute under multiplication, they are linearly independent), so $\underline{e}, \underline{i}, \underline{j}, \underline{k}$ form a basis for V. The function $f : \mathbb{H} \to U$ given by

$$f(x_0 + x_1 i + x_2 j + x_3 k) = x_0\underline{e} + x_1\underline{i} + x_2\underline{j} + x_3\underline{k}$$

for $x_r \in \mathbb{R}$ is a skew-field isomorphism.   ∎

Of course, we have not shown that it is impossible to generalise some aspects of the quaternions, but merely that we cannot generalise them all simultaneously.

Exercise D.10 *Consider our proof of Theorem D.4*

  (i) *Check that D^- and D^+ are vector subspaces of V.*
  (ii) *Check that T is linear.*
  (iii) *Show that, if $\underline{p} \in D^-$, then $\underline{p}^{-1} \in D^-$.*
  (iv) *Check that S takes D^- to D^+.*
  (v) *Check the values of $\underline{k} \otimes \underline{j}$, $\underline{k} \otimes \underline{i}$ and $\underline{i} \otimes \underline{k}$.*
  (vi) *Check, in as much detail as you consider necessary, that f is a skew-field isomorphism.*

References

[1] N. Ashby. Relativity in the global positioning system. *Living Reviews in Relativity*, 6, 2003.

[2] W. W. Rouse Ball. *History of the Study of Mathematics at Cambridge*. Cambridge University Press, New York, 1889. There is a Cambridge Library Collection reprint.

[3] E. R. Berlekamp, J. H. Conway and R. K. Guy. *Winning Ways for Your Mathematical Plays*. Academic Press, Cambridge, Mass., 1982. Two volumes.

[4] R. P. Boas. *Lion Hunting and Other Mathematical Pursuits*, volume 15 of *Dolciani Mathematical Expositions*. Mathematical Association of America, Washington, D.C., 1995.

[5] H. T. Colebrooke. *Algebra, with Arithmetic and Mensuration, from the Sanskrit of Brahmegupta and Bháscara*. John Murray, London, 1817. Reissued by Cambridge University Press.

[6] M. J. Crowe. *A History of Vector Analysis*. University of Notre Dame Press, Notre Dame, Ind., 1967. Reissued by Dover.

[7] R. Dedekind. *Essays on the Theory of Numbers*. Open Court Publishing Company, New York, 1901. Translation by W. W. Beman. This was reissued by Dover and is also available from Project Gutenberg.

[8] Euclid. *Euclid's Elements*. Cambridge University Press, Cambridge, 2nd edn, 1926. Translation by T. L. Heath. Reissued by Dover.

[9] H. M. Friedman. Philosophy 532 and Philosophy 536. https://u.osu.edu/friedman.8/files/2014/01/Princeton532-1pa84c4.pdf, 2002.

[10] P. Gordon. Numerical cognition without words: Evidence from Amazonia. *Science*, 306:496–499, 2004.

[11] R. P. Graves. *Life of Sir William Rowan Hamilton*. Hodges, Figgis, Dublin, 1882–1889. Three volumes.

[12] P. R. Halmos. *Naive Set Theory*. Van Nostrand, Princeton, N.J., 1960. The later publishing history is complicated, but the book is available in many libraries and is usually in print.

[13] G. H. Hardy. *A Mathematician's Apology*. Cambridge University Press, Cambridge, 1940.

[14] L. Hogben. *Mathematics for the Million*. Norton, New York, 1937.

[15] A. Imhausen. *Mathematics in Ancient Egypt.* Princeton University Press, Princeton, N.J., 2016.

[16] P. T. Johnstone. *Notes on Logic and Set Theory.* Cambridge University Press, Cambridge, 1987.

[17] H. C. Kennedy. Peano. In *Dictionary of Scientific Biography.* Charles Scribner's Sons, New York, 1970–1980.

[18] H. Ketcham. *The Life of Abraham Lincoln.* A. L. Burt, New York, 1901. Available at Project Gutenberg.

[19] S. Körner. *The Philosophy of Mathematics.* Hutchinson, University Library, London, 1960. Reissued by Dover.

[20] E. Landau. *Foundations of Analysis.* Chelsea, New York, 1951. Translation by F. Steinhadt of *Grundlagen der Analysis,* Akademische Verlagsgesellschaft, Leipzig, 1930.

[21] C. E. Linderholm. *Mathematics Made Difficult.* Wolfe Publishing, London, 1971.

[22] J. E. Littlewood. *Littlewood's Miscellany.* Cambridge University Press, Cambridge, 1986. Second edition of *A Mathematician's Miscellany,* edited by B. Bollobás.

[23] T. Mills. How fast will future warming be? www.thegwpf.org/statistical-forecasting-how-fast-will-future-warming-be/, 2016.

[24] E. Nagel and J. R. Newman. *Gödel's Proof.* New York University Press, New York, 1958.

[25] R. P. Nederpelt, J. H. Geuvers and R. C. de Vrijer, eds. *Selected Papers on Automath.* Elsevier, Amsterdam, 1994.

[26] A. Pais. *Subtle Is the Lord.* Oxford University Press, Oxford, 1982.

[27] R. S. Palais. The classification of real division algebras. *American Mathematical Monthly,* 75:366–368, 1968.

[28] B. Russell. *Mysticism and Logic and Other Essays.* Allen and Unwin, London, 1917. Available at Project Gutenberg.

[29] D. E. Smith. *A Source Book in Mathematics.* McGraw-Hill, New York, 1929. Reissued by Dover.

[30] M. Spivak. *Calculus.* Cambridge University Press, Cambridge, 3rd edn, 1994.

[31] J. van Heijenoort. *From Frege to Gödel.* Harvard University Press, Cambridge, Mass., 1967.

[32] R. Westfall. *Never at Rest.* Cambridge University Press, Cambridge, 1980.

Index

Printed in the United States
By Bookmasters